Criteria for Divisibility

T0138392

Popular Lectures in Mathematics

Survey of Recent East European Mathematical Literature

A project conducted by
IZAAK WIRSZUP,
Department of Mathematics,
the University of Chicago,
under a grant from the
National Science Foundation

N. N. Vorob'ev

Criteria for Divisibility

Translated and
adapted from the
Russian edition by
Daniel A. Levine and
Timothy McLarnan

The
University of Chicago
Press
Chicago and
London

The University of Chicago Press, Chicago 60637
The University of Chicago Press, Ltd., London

N. N. VOROB'EV is a senior researcher at the
Leningrad Division of the Mathematics Institute
of the USSR Academy of Sciences

Library of Congress Cataloging in Publication Data

Vorob'ev, Nikolaĭ Nikolaevich, 1925–
   Criteria for divisibility.

   (Survey of recent East European mathematical
literature) (Popular lectures in mathematics)
   Translation of Priznaki delimosti.
   1. Division. I. Title.
QA242.V6713   513′.2   74-11634
ISBN 0-226-86516-9

Contents

Preface

(to which the author advises the reader
to give his special attention)

Modern mathematics education in the schools is oriented mainly toward training the student to think in terms of functional relations and to deal with continuous mathematical objects. The changes planned in the school curricula [of the USSR] do not affect this orientation. The last few years have seen the intensive development of new areas of applied mathematics: computer programming, some aspects of cybernetics and operations research, mathematical economics, linguistics, and so on. A complete understanding of these new sciences and of more classical studies, however, requires the development of new and fruitful abstractions for dealing with discrete entities and with combinatorial problems. There is, therefore, a need for a treatment in the popular scientific literature of these aspects of mathematics.

Mathematics, like a forest, can be approached and explored by many paths, and in any one investigation only a small sample of its riches can be examined. This book may be regarded as a description of one of the possible walks along the edge of modern mathematics. We shall use our investigation of criteria for divisibility as a framework within which we can develop some of the abstract notions of discrete mathematics. First, ideas arising from the basic theorems of elementary number theory (such as the fundamental theorem of arithmetic) will be considered. On a more abstract level, divisibility itself will then be considered as a relation on the set of integers; that is, as one example of a more general notion. Finally, criteria for divisibility as such will be treated as algorithms which enable us to tell whether a given number is divisible by another. Criteria for congruence, algorithms which act on any number to produce its remainder on division by a given number, will also be considered separately.

A number of assertions in this book will be established by two different methods in order to demonstrate the multitude of interrelations

vii

between individual mathematical facts and the possibility of various approaches to the same subject.

This book is intended for mathematically inclined students in the upper grades of secondary school. It presupposes no preliminary knowledge except the ability to manipulate algebraic identities and a familiarity with the binomial formula. However, since the logical structure of the material is fairly complicated, its mastery may require much attention and patience.

The following plan for the study of this book is recommended. The first reading may be limited to the basic text of chapters 1–3, and all problems except numbers 3.3, 3.5, 3.7, 3.13, 3.15, 3.17, and 3.18 may be ignored. This will provide a general, descriptive acquaintance with the subject. Most students who are inexperienced in mathematics seem convinced of the validity of the theorem on the unique decomposition of a natural number into prime factors (apparently considering it almost as an axiom), and they may therefore simply accept it as true and regard theorems 1.9–1.13 as its corollaries. On the second reading, however, the student should provide independent proofs of all theorems in the order in which they are given. In order that the reader not succumb too often to the temptation to use prepared proofs, all of the proofs of these theorems have been placed in a special section. The sole exception is the proof of theorem 1.7, which has been selected as an example to illustrate to the reader the desired level of rigor. Chapter 4 should also be studied on the second reading, and the problems of the basic text should be solved. Finally, on the third reading, the sections in small print should be read and the problems relating to them should be solved.[1]

The reader who wishes to deepen his knowledge of the theory of numbers should consult the classic work of Professor I. M. Vinogradov, *Elements of Number Theory* (New York: Dover, 1954), translated from a later edition as *An Introduction to the Theory of Numbers* (London and New York: Pergamon Press, 1955). For the abstract theory of relations on a set we recommend *Lectures on General Algebra* by A. G. Kurosh (Oxford and New York: Pergamon Press, 1965) or *Lattice Theory* by Garrett Birkhoff (Providence, R.I.: American Mathematical Society, 1967). Finally, a more detailed and systematic explanation of the concept of an algorithm may be found in the booklet by B. A. Trakhtenbrot, *Algorithms and Automatic Computing Machines* (Boston: D. C. Heath, 1963); and a rigorous treatment of the theory of algorithms is given in

---

1. Since the average Soviet student has perhaps had more experience with rigorous mathematical proof than his western counterpart, these directions are particularly important to American students.

a monograph by A. A. Markov, one of the founders of this theory: *The Theory of Algorithms* (Jerusalem: Program for Scientific Translations, 1961). [Available from the Office of Technical Services, U.S. Department of Commerce, Washington, D.C.]

It is by no means correct to suppose that all of these theorems are obvious and require no proof. One must remember that in mathematics any statement other than an axiom or a definition requires proof. It is necessary to prove the facts listed above (for example, the reflexive law of divisibility), because we cannot derive them from the definition of divisibility alone—we must also use the properties of the integers.

The following example will help to make this clear. The sum, difference, and product of any two even numbers is always an even number. It is not always possible, however, to divide a given even number by another. Moreover, even if the division is possible, the quotient need not be even. Therefore, we can introduce the concept of the "even divisibility" of even numbers.

DEFINITION 1.2. *The even number a is* evenly divisible *by the even number b if there exists an even number c such that a = bc.*

It is clear that theorem 1.1 is false for even divisibility since for any nonzero a there exists no even number c such that $a = ac$.

We shall consider more questions concerning even divisibility later. The example of even divisibility shows that it is possible to construct theories of divisibility on sets other than the set of integers (in this case, the set of even integers) in such a way that theorems which are true for some of these theories are false for others.

Problems. Prove the following assertions:

Problem 1.1. $a|0$.

Problem 1.2. $1|a$.

Problem 1.3. If $a|1$, then $a = \pm 1$.

Problem 1.4. For any $a \neq 0$, there exists a number b different from a such that $a|b$.

Problem 1.5. For any number a there exists a number b different from a such that from $c|b$ and $a|c$ it follows that $c = b$ or $c = a$ (a, b, and c are assumed to be natural numbers).

Problem 1.6. Prove the analogues of theorems 1.2, 1.3, 1.4, and 1.5 for the case of even divisibility.

Problem 1.7. Construct a theory of divisibility in which theorems 1.1, 1.3, and 1.4 are true, but not theorems 1.2 and 1.6.

1.3

The most cursory acquaintance with the concrete facts of divisibility convinces us that the divisibility of numbers is in practice not strongly related to their magnitude. One may find fairly small numbers that have a comparatively large number of divisors. For example, 12 is divisible by 1, 2, 3, 4, 6, and 12; and the number 60 has 12 divisors. These numbers so rich in divisors may be contrasted, however, with extremely large numbers that have the minimal number of divisors— two (by theorem 1.1 and problem 1.2 every number distinct from 1 has at least two different positive divisors). In fact, some laws connecting the divisibility properties of numbers and their magnitude are known, but these laws are so complicated and intricate that we shall not touch upon them here.

1.4

It is perhaps more interesting that divisibility itself allows us to establish a certain ordering of the integers that is different from the usual ordering, but has much in common with it.

Indeed, let us consider what we mean by the possibility of ordering the integers by their magnitude. It is not hard to see that by this possibility we mean that, for some pairs of numbers a and b, the relation "greater than or equal to" holds:

$$a \geq b,$$

which signifies that $a - b$ is nonnegative (that is, there must exist a natural number c such that $a = b + c$). But the notion of divisibility is nothing more than the idea that some pairs of numbers a and b satisfy some other mathematical relationship (namely, that there exists a number c such that $a = bc$). Thus, the relation of divisibility and the relation "greater than or equal to" are concepts of the same nature, and thus we may speak of their common properties, or, on the other hand, of their differences.

We note immediately that the relation "greater than or equal to" has more properties in common with the relation of divisibility than does the relation "greater than." This is partly because the relation "greater than or equal to," like the relation of divisibility, is reflexive (that is, the relation $a \geq a$ is true for any a), while the relation "greater than" is not reflexive (indeed, the inequality $a > a$ is never valid). It is precisely because of this that the relation "greater than or equal to" is considered to be an order relation on the natural numbers, while the apparently simpler and more natural relation "greater than" is not.

1.5

The relation $\geq$ has the following easily verifiable properties:

Property 1.1. $a \geq a$ (reflexivity).

Property 1.2. If $a \geq b$ and $b \geq a$, then $a = b$ (antisymmetry).

Property 1.3. If $a \geq b$ and $b \geq c$, then $a \geq c$ (transitivity).

Property 1.4. In any sequence of natural numbers with distinct terms,

$$a_1 \geq a_2 \geq a_3 \geq \cdots \geq a_n \geq \cdots,$$

there is a "last number." This property of the relation is sometimes called the *well-ordering principle* for the set of natural numbers.

The well-ordering principle is fairly complicated and might seem at first to be a bit artificial, but it reveals extremely important features of the structure of the set of natural numbers ordered by the relation $\geq$, and many other properties of this relation may be derived from it. Moreover, we shall discover that it is the basis for the method of "proof by induction," which is widely used in mathematical arguments.

As a useful application of this property, we shall prove that there exists a natural number a such that for any natural number b, if $a \geq b$, then $a = b$.

Indeed, if there were no such number, then for each a_n we could find an a_{n+1} such that $a_n \geq a_{n+1}$ and $a_n \neq a_{n+1}$. Starting with an arbitrary a_1, we could then obtain a sequence with distinct terms

$$a_1 \geq a_2 \geq a_3 \geq \cdots \geq a_n \geq a_{n+1} \geq \cdots$$

which never ends. But the existence of such a sequence contradicts the well-ordering principle for the natural numbers.

Thus, the existence of the number a mentioned above has been proved. This number is called the *first* or *minimal* natural number (clearly, it is the number zero). We note, however, that we have not yet established the uniqueness of the minimal number. This uniqueness must be shown by a more roundabout method.

Property 1.5. For any number a there exists a number b distinct from a such that $b \geq a$.

This property of the set of natural numbers is called its *unboundedness* under the relation $\geq$.

Property 1.6. For any number a that is not minimal, there exists a number b such that $a \geq b$, $a \neq b$ and such that for any number c, $a \geq c \geq b$ implies that $c = a$ or $c = b$. This formal assertion, when translated into more descriptive language, means that every natural number except 0 has an immediate predecessor. (This could be formulated differently by saying that the set of all numbers less than a given number has a largest element.)

Property 1.7. Either $a \geq b$ or $b \geq a$. This property of the relation $\geq$ is called the *law of dichotomy*. In mathematics the term "dichotomy" is used to express the necessity of having one of two possibilities realized. The word itself is of Greek origin and means "a division into two parts."

We emphasize that properties 1.1 through 1.7 are properties of the *relation* $\geq$ on the natural numbers and are not properties of the individual numbers connected by this relation. Thus, some of the properties 1.1 through 1.7 may not be true of relations other than the magnitude relation $\geq$.

Problem 1.8. Using only properties 1.1 through 1.7 of the ordering $\geq$ and no other properties of the natural numbers themselves or of the operations on them,

  a. prove the uniqueness of the minimal element;

  b. prove the uniqueness of the immediate predecessor;

  c. formulate the definition of the immediate successor of a given number a (that is, of the number $a + 1$), and prove its existence and uniqueness.

Problem 1.9. Determine which of the assertions (properties) 1.1 through 1.7 remain valid for the relation "greater than" ($>$).

1.6

The validity of properties of the relation $\geq$ (and of properties of any other relation as well) may be established in either of two ways. First, we may use properties of individual numbers or known peculiarities of the structure of the set of all natural numbers. This is precisely the way in which properties 1.1 through 1.7 were verified. Then, having convinced ourselves that properties 1.1 through 1.7 are correct, we may begin to reason abstractly and may derive further properties of this relation solely from properties 1.1 through 1.7. This is the way in which we proved the existence of the minimal element and the assertions of problem 1.8.

A second approach to this problem is particularly widely used in modern mathematics. This is the so-called axiomatic approach in which certain *axioms* are laid down (in our case, properties 1.1 through 1.7) which reflect the basic properties of the objects studied and are not subject to proof. All further assertions, called *theorems*, are derived from these axioms by strictly logical means, without any reference to the objects studied.

It may seem to some readers that the consideration of relations in isolation from the objects connected by these relations (for example, numbers) is the height of mathematical abstraction, and that so much abstraction is unnecessary in practical life. We shall make two remarks on this point.

In the first place, the arguments presented here are not especially "abstract" from the standpoint of modern mathematics. Contemporary mathematicians, moreover, find it necessary to consider many relations simultaneously, and even to connect pairs of relations by new relations (by "relations of the second order," so to speak).

The material presented so far will enable us to illustrate the concept of a relation between relations by an example:

Let $\alpha, \beta, \ldots$ be some collection of relations "connecting" the natural numbers. By this we mean that for any pair of numbers a and b and any relation γ of our collection, we know whether or not the numbers a and b are related (or connected) by γ. If a and b are related by γ, we shall write $a\gamma b$.

We shall say that the relation α is *stronger* than the relation β and write $\alpha \supset \beta$ if any pair of numbers connected by the relation β is also connected by the relation α; that is, if $a\beta b$ implies $a\alpha b$.

Thus, denoting the relation of even divisibility by $|_e$, we may write $| \supset |_e$. Furthermore, it is clear that $\geq \supset >$. On the other hand, neither $| \supset \geq$ nor $\geq \supset |$ holds: The relations of divisibility and of "greater than or equal to" are not connected by the second-order relation $\supset$. It is precisely this situation that was described in words in section 1.3.

Further training is, of course, necessary in order to operate freely with such complicated concepts as relations between relations.

In the second place, similar and even more abstract reasoning is being employed more and more frequently in the applications of mathematics to economics, biology, linguistics, and military affairs. Unfortunately, a more detailed explanation of this reasoning and of the reasons for its use would take us too far from our basic topic.

1.7

The method of proof known as *mathematical induction* (sometimes called *complete induction*) is intimately tied to the ordering of the natural numbers by the magnitude relation $\geq$. This method is usually applied in the following form:

Let $A(n)$ be some assertion concerning an arbitrary natural number n. In order to prove that $A(n)$ is true for any natural number n, we must establish the infinite sequence of assertions

$$A(0), A(1), \ldots, A(n), \ldots .$$

We shall suppose that:

a. the assertion $A(0)$ is valid (the "basis for induction");[2]

b. from the validity of the assertion $A(n)$, there follows the validity of the assertion $A(n + 1)$ (the "inductive step").

The principle of mathematical induction asserts that under the hypotheses (a) and (b), $A(n)$ is valid for any natural number n.

The principle of mathematical induction is not an independent statement, but may be derived from properties 1.1 through 1.7 of the ordering of the set of natural numbers under $\geq$.

Indeed, let us suppose that conditions (a) and (b) are satisfied for the assertions $A(n)$, but that the conclusion asserted by induction does not hold. This means that there must exist numbers m for which the assertion $A(m)$ is false. Let m_1 be one such number. If for all $n < m_1$ the assertion $A(n)$ is true, then m_1 is the least of the numbers for which $A(n)$ does not hold. But if $A(n)$ is true for not all $n < m_1$, then there must exist an $m_2 < m_1$ such that $A(m_2)$ is false.

As a result, we arrive at a sequence of distinct natural numbers,

$$m_1 \geq m_2 \geq \cdots \geq m_r \geq \cdots , \tag{1.2}$$

2. The assertion $A(1)$ is often taken as the basis for induction. Clearly, this difference is not important. It is only important that the basis of induction involve the first of the numbers we consider.

for which $A(n)$ does not hold. By the well-ordering principle (property 1.4), there must be a last element m_r in the sequence (1.2). It is clear that the number m_r is the smallest number for which $A(n)$ is false.

Since $A(0)$ is true by hypothesis, $m_r \neq 0$, and so there exists a number m_r^* immediately preceding m_r (in actuality this number is $m_r - 1$). Since $m_r^* < m_r$, the assertion $A(m_r^*)$ must be true. But then by hypothesis (b) of the principle of mathematical induction, the assertion $A(m_r^* + 1)$, that is, the assertion $A(m_r)$, must also be true, and we have obtained a contradiction. This contradiction shows that there can be no natural numbers m for which $A(m)$ does not hold.

The above argument should not be taken as a proof of the absolute truth of the principle of mathematical induction. It shows only that it is possible to derive this one mathematical assertion (the validity of the method of induction) from others (the properties of the relation $\geq$). We took these properties as axioms and, accordingly, did not prove them but only verified them as being consistent with our intuitions. Any attempt to give a mathematical proof for them would require the introduction of new assertions as axioms.

In particular, proofs of the well-ordering principle necessarily involve some sort of inductive argument (the reader can convince himself of this independently).

The booklets by I. S. Sominskii, *The Method of Mathematical Induction*, and by L. I. Golovina and I. M. Yaglom, *Induction in Geometry* (Boston: D.C. Heath, 1963), contain a large number of examples of the use of this method. Inductive proofs will be used frequently in this book.

Problem 1.10. Let pairs of objects of an arbitrary nature (for example, numbers, points, functions, theorems, and so on) be connected by some relation $\varepsilon-$ satisfying properties 1.1 through 1.7. Prove that these objects (elements) can be enumerated (that is, written in some order) $A_1, A_2, \ldots, A_n, \ldots$ in such a way that $A_i \varepsilon- A_j$ if and only if $i \geq j$.

In essence, the statement above means that a relation having properties 1.1 through 1.7 orders the set on which it is defined in a linear chain of elements:

$$A_1 \varepsilon- A_2 \varepsilon- A_3 \varepsilon- \cdots .$$

1.8

Let us return, however, to the relation of divisibility. In the case of the positive integers, theorems 1.1, 1.2, and 1.3, and problems 1.3, 1.4, and 1.5 show that in properties 1.1 through 1.6 we may replace the relation $\geq$ by the relation $|$. The statement of property 1.7 regarding divisibility would assert that "given any two positive integers, at least one is divisible by the other," but this is clearly false. Thus, the divisibility relation has all but one of the properties of the magnitude relation. As a result, the relation of divisibility orders the natural numbers, but in a more complicated manner than that of a linear chain (see figure 1.1). We note that numbers close to each other in terms of magnitude often turn out to be fairly "far" from each other in the sense of divisibility. The numbers 4 and 5 or 7 and 8 demonstrate this clearly.

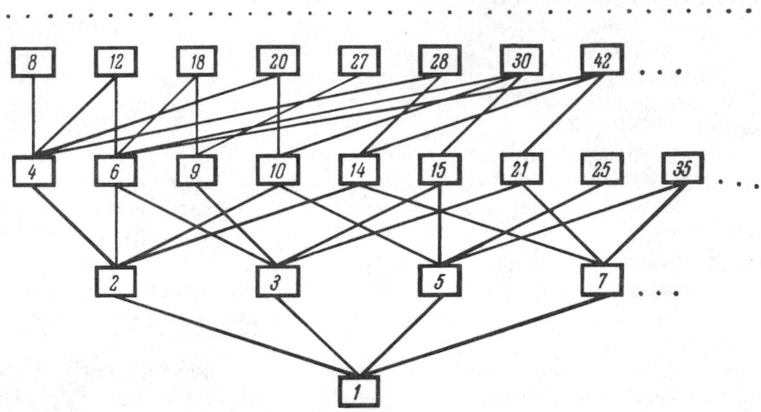

Fig. 1.1

Let us try to pass from the divisibility of positive integers to the divisibility of the natural numbers; that is, let us try to include zero in our considerations. Then the scheme in the figure would be enlarged to include a box lying above all the other boxes in the scheme, for zero is divisible by any natural number, while no natural number distinct from zero is divisible by zero.

We leave it to the reader to reformulate and verify assertions 1.1 through 1.7 for the case in which zero is included.

1.9

DEFINITION 1.3. *Any relation $\vdash$ satisfying the conditions*
(1.1′) reflexivity ($a \vdash a$),
(1.2′) antisymmetry (from $a \vdash b$ and $b \vdash a$ it follows that $a = b$), and
(1.3′) transitivity (from $a \vdash b$ and $b \vdash c$ it follows that $a \vdash c$),
is called a relation of partial ordering.

Relations of partial ordering play an important role in situations in which there is no "natural" linear ordering, such as situations in which each object is identified by several indices that are qualitatively not comparable to one another.

For example, consider the results of a sports competition that includes several different sports. If one team captures higher places than another in all of the sports represented, it would be natural to consider that the first team has attained greater distinction. But if the first team captures higher places in all events except croquet, in which the second team is stronger, then the question of the final relative standing of the teams would not be so clear. Croquet enthusiasts might even insist on giving a higher place to the second team. In any event, any final distribution of places must be connected with some agreement on how to evaluate the points.

1.10

It is easy to construct relations satisfying conditions 1.1' through 1.3' of definition 1.3; that is, relations of partial ordering. Indeed, a great many types of objects can be partially ordered, often in several different ways. As a consequence, there is very little that can be said about a general partial ordering except that it satisfies our three axioms. In particular, we cannot always apply the method of mathematical induction to the objects on which a partial ordering is defined.

Suppose, however, that the following conditions are added to conditions 1.1' through 1.3': (1.4') well-ordering; (1.5') unboundedness; (1.6') each object distinct from the minimal one has at least one immediate predecessor; and two new conditions:

Property 1.8. Each object has only finitely many predecessors;

Property 1.9. For any a and $b \xleftarrow{} a$ ($b \neq a$) there exists a c preceding b and such that $c \xleftarrow{} a$.

It turns out that on the basis of the partial ordering of the set of natural numbers by a relation that satisfies conditions 1.1' through 1.6', 1.8, and 1.9, it is possible to construct a modification of the method of mathematical induction. This consists of the following:

Suppose, once again, that $A(n)$ is some assertion about an arbitrary object n in our given set. We shall assume that

a. the assertion $A(a)$ is valid, where a is any minimal object under the ordering $\xleftarrow{}$;

b. if n is any object, and all assertions of the form $A(m)$, where $n \xleftarrow{} m$ and $n \neq m$, are valid, then $A(n)$ is also valid.

The new form of the principle of induction states that if conditions (a) and (b) are fulfilled, then $A(n)$ is true for any object n in our set.

Problem 1.11. Derive the "new form" of the principle of induction from its "old form."

Since the relation of divisibility on the natural numbers satisfies conditions 1.1' through 1.6', 1.8, and 1.9 (formulate and check conditions 1.8 and 1.9 for the relation of divisibility), this principle of induction is applicable to the relation of divisibility.

As applied to divisibility, the new principle of induction may be formulated as follows: If some assertion $A(n)$ is valid for $n = 1$, and its validity for all divisors of a number n distinct from n implies its validity for n, then it holds for any n.

1.11

Since the division of integers (as we have seen) is not always possible, it is natural to consider—along with division—another more general operation that may always be performed and which coincides with division when the latter is possible. This operation is called *division with remainder*.

DEFINITION 1.4. *To* divide *a number a by a number b ($b > 0$) with remainder* means *to represent the number a in the form*

$$a = bq + r \, ,$$

where q and r are integers and $0 \leq r < b$.

The number q so obtained is called the *partial quotient* and the number r the *remainder*. Clearly, $r = 0$ if and only if $b|a$. In this case q is equal to the quotient of the division of a by b.

We shall show that division with remainder may always be performed and that the partial quotient and the remainder are completely determined by the dividend and divisor—that is, that they are unique.

Suppose first that $a \geq 0$. We write the numbers

$$a, a - b, a - 2b, \ldots , \qquad (1.3)$$

one after another until a negative number appears. Clearly, such a number must eventually appear.

(In fact, this follows from the well-ordering of the set of natural numbers by the relation $\geq$.)

Suppose that the last of the nonnegative elements of sequence (1.3), that is, the smallest of them, is the number $a - bq$. Denoting this number by r, we have

$$a = bq + r \, . \qquad (1.4)$$

Clearly, $r < b$ (otherwise the number $r - b$, that is, $a - (q + 1)b$, would be nonnegative, and this cannot be, since r is the *least* of the nonnegative numbers in (1.3)). Thus, (1.4) is the desired representation of the number a.

Now suppose that $a < 0$. Reasoning in a manner similar to the above argument, we write the sequence of numbers

$$a, a + b, a + 2b, \ldots , $$

until the first nonnegative number r appears (it is easy to check that $r < b$). Let

$$r = a + bq' \, . $$

Then, denoting $-q'$ by q, we obtain

$$a = bq + r,$$

as was required.

The possibility of division with remainder is thus proved in all cases.

We shall now prove the uniqueness of the partial quotient and remainder; that is, we shall prove that if

$$a = bq + r, \tag{1.5}$$

and, at the same time,

$$a = bq_1 + r_1, \tag{1.6}$$

with $0 \leq r < b$ and $0 \leq r_1 < b$, then $q = q_1$ and $r = r_1$.

In order to prove uniqueness, it does not suffice simply to note that the numbers q and r obtained by the process above are completely determined by a and b, since such an argument in no way precludes the possibility that other values of q and r may be obtained by some completely different method.

To give a complete proof of uniqueness, therefore, we combine equations (1.5) and (1.6) to obtain

$$bq + r = bq_1 + r_1,$$

that is,

$$r - r_1 = b(q_1 - q);$$

which says that $r - r_1$ is divisible by b. But $|r - r_1| < b$ (since $0 \leq r < b$ and $0 \leq r_1 < b$), and, according to theorem 1.4, this is possible only if $r - r_1 = 0$; that is, if $r = r_1$. But then

$$b(q_1 - q) = 0,$$

and since b is not zero, $q_1 - q = 0$; that is, $q_1 = q$.

Thus we have proved the following theorem.

THEOREM 1.7 (Division with remainder). *Given any numbers a and b ($b > 0$), there exist unique numbers r and q such that $a = bq + r$, and $0 \leq r < b$.*

Problem 1.12. Formulate and prove a theorem on division with remainder for even divisibility.

1.12

DEFINITION 1.5. *A positive integer p not equal to* 1 *is said to be* prime *if it is divisible by itself and by* 1, *and by no other positive integers.*

Examples of prime numbers are 2, 3, 5, 7, 11, and 13.

A positive integer distinct from 1 which is not prime is said to be *composite.*

THEOREM 1.8. *There exist infinitely many prime numbers.*

Any number that divides both the numbers a and b is called a *common divisor* of these numbers. The largest of the common divisors of a and b is called their *greatest common divisor* and is denoted by (a, b).

If the greatest common divisor of the numbers a and b is 1, then these numbers are said to be *relatively prime.*

In other words, the numbers a and b are relatively prime if they are not both divisible by any positive integer except 1.

THEOREM 1.9. *If a and p are natural numbers and p is a prime, then either $p|a$ or the numbers a and p are relatively prime.*

Any number that is divisible by both a and b is called a *common multiple* of these numbers. The smallest positive common multiple of a and b is called their *least common multiple.*

THEOREM 1.10. *If M is a common multiple of a and b and m is their least common multiple, then $m|M$.*

THEOREM 1.11. *The least common multiple of two relatively prime numbers is their product.*

COROLLARY. *A number a is divisible by the relatively prime numbers b and c if and only if it is divisible by their product bc.*

THEOREM 1.12. *If $c|ab$ and b and c are relatively prime, then $c|a$.*

THEOREM 1.13. *If the product of several factors is divisible by a prime p, then at least one of the factors is divisible by p.*

COROLLARY. *If p is a prime and $0 < k < p$, then the number*

$$\binom{p}{k} = \frac{1 \cdot 2 \cdots (p-1)p}{1 \cdot 2 \cdots (k-1)k \cdot 1 \cdot 2 \cdots (p-k-1)(p-k)}$$

is divisible by p.

THEOREM 1.14 (The fundamental theorem of arithmetic). *Any positive integer other than* 1 *can be represented as a product of prime factors. This representation is, moreover, unique* (if products that differ only in the order of factors are not considered distinct).

The fundamental theorem of arithmetic asserts the possibility, in principle, of decomposing any number into prime factors. The actual construction of such a decomposition, however, is often extremely difficult, especially if the numbers involved are very large. Often, large numbers can be factored or proved to be prime only with the aid of modern electronic computers. It was not until 1957, for example, that the number $2^{3217} - 1$ was shown to be prime. This number has 969 digits and is the largest known prime at this time (1963). To prove that it is prime required five and one-half hours of computation by a large computer.[3]

Suppose that some number a is decomposed into a product of prime factors. Collecting equal factors, we may obtain a formula of the form

$$a = p_1^{\alpha_1} p_2^{\alpha_2} \cdots p_r^{\alpha_r}, \qquad (1.7)$$

where $p_1, p_2, \ldots, p_r$ are distinct and $\alpha_1, \alpha_2, \ldots, \alpha_r$ are positive numbers. The product on the right-hand side of formula (1.7) is called the *canonical decomposition* of the number a.

THEOREM 1.15. *A necessary and sufficient condition for the numbers a and b to be relatively prime is that no prime factor appearing in the canonical decomposition of a also appears in the canonical decomposition of b.*

THEOREM 1.16. *Let* (1.7) *be the canonical decomposition of the number a. Then $a|b$ if and only if*

$$p_1^{\alpha_1}|b, \ p_2^{\alpha_2}|b, \ldots, p_r^{\alpha_r}|b .$$

From theorems 1.15 and 1.16 it follows that divisibility by a product of several relatively prime numbers is equivalent to simultaneous divisibility by each of them.

Problem 1.13. Determine an upper bound for the least prime divisor of a composite number a.

Problem 1.14. Let

$$a = p_1^{\alpha_1} p_2^{\alpha_2} \cdots p_r^{\alpha_r}$$

3. The prime $2^{19937} - 1$ was discovered in 1972. It has 5984 digits.

16 Chapter One

be the canonical decomposition of the number a. Then a necessary and sufficient condition for $b \mid a$ is that the canonical decomposition of b have the form

$$b = p_1^{\beta_1} p_2^{\beta_2} \cdots p_r^{\beta_r},$$

where $0 \le \beta_1 \le \alpha_1,\, 0 \le \beta_2 \le \alpha_2, \ldots, 0 \le \beta_r \le \alpha_r$.

Problem 1.15. Let us denote by $\tau(a)$ the number of distinct positive divisors of the number a (including 1 and the number a itself). Show that for a number a with the canonical decomposition $p_1^{\alpha_1} p_2^{\alpha_2} \cdots p_r^{\alpha_r}$,

$$\tau(a) = (\alpha_1 + 1)(\alpha_2 + 1) \cdots (\alpha_r + 1).$$

Problem 1.16. Find a if it is known that $3 \mid a$, $4 \mid a$, and $\tau(a) = 14$.

Problem 1.17. The canonical decomposition of the number a has the form $p_1^{\alpha_1} p_2^{\alpha_2}$, with $\alpha_1 > 0$, $\alpha_2 > 0$, and $\tau(a^2) = 81$. What are the possible values of $\tau(a^3)$?

Problem 1.18. What is a if $a = 2\tau(a)$?

Problem 1.19. Are the analogues of theorems 1.11–1.14 true for even divisibility?

Problem 1.20. Find a method of constructing the canonical decomposition of the least common multiple and greatest common divisor of two numbers whose canonical decompositions are given.

2   The Divisibility of Sums and Products

2.1

In many cases, only the remainder in a problem of division with remainder is of interest, and the value of the partial quotient is of no importance.

Suppose that we want to know on what day of the week 1 January 2000 will fall (provided, of course, that we keep the same calendar we use now). By consulting a calendar, it is easy to find that January 1, 1974, fell on a Tuesday. The twenty-six years that separate these dates contain $26 \cdot 365 + 6$ days (the last term being the number of leap years during this period), that is, 9496 days. These days amount to 1356 weeks with four days left over. After 1356 weeks pass, it will again be Tuesday, so that after another four days, on 1 January 2000, it will be Saturday. It is clear that, for the solution of the problem we have just considered, it is totally unnecessary to know just how many whole weeks pass in the 26 years, and that all that matters is the number of days left over after these weeks.

This example is not completely useless, for historians, especially those who study the Orient, must often consider such problems in comparing dates given in various calendars.

It might seem at first that the simplest method of finding the remainder on division of one number by another is to perform the division with remainder directly. In practice, however, performing such a division is often extremely tedious, especially if the dividend is not written in the decimal system to which we are so accustomed, but instead is given by some complicated expression such as $2^{1000} + 3^{1000}$. In addition, the lion's share of the work is wasted in finding the partial quotient, for which we have no need. It is therefore necessary to look for a method of determining the remainder directly, without calculating the partial quotient.

We shall demonstrate one such method on the problem that we have just solved concerning the date January 1, 2000. We may argue in the following manner: Each year that is not a leap year consists of 365 days, which fill out 52 weeks with one day left over. A leap year consists of the same number of weeks with two days left over. This means that the period from January 1, 1974, to January 1, 2000, consists of some (wholly unimportant) number of complete weeks plus a number of days that is equal to the number of years in this period, with each leap year counted twice. This number of days is equal to $26 + 6 = 32$. Taking four complete weeks away from these days, we have four days left, which we proceed to count off as before, beginning with Tuesday. It can be seen that this "replacement" of years by days is an instance of an extremely general method, which we shall now begin to study.

2.2

DEFINITION 2.1. *We say that the numbers a and b are* congruent modulo m *if the remainders on division of a and b by m are equal.*

We shall now establish some properties of congruent numbers.

THEOREM 2.1. *The numbers a and b are congruent modulo m if and only if $m|a - b$.*

THEOREM 2.2. *If the numbers $a_1, a_2, \ldots, a_n$ are congruent modulo m to $b_1, b_2, \ldots, b_n$, respectively, then the numbers $a_1 + a_2 + \cdots + a_n$ and $b_1 + b_2 + \cdots + b_n$, as well as $a_1 a_2 \cdots a_n$ and $b_1 b_2 \cdots b_n$, are also congruent modulo m.*

COROLLARY. *If the numbers a and b are congruent modulo m, then so are a^n and b^n for any natural number n.*

Theorem 2.2 and its consequences provide fairly rich possibilities for finding remainders on division. We shall present some examples:

Example 2.1. Find the remainder on division by 3 of the number

$$A = 13^{16} - 2^{25} \cdot 5^{15}.$$

Clearly, the number 13 is congruent to 1, 2 congruent to -1, and 5 also congruent to -1 modulo 3. So, on the basis of the discussion above, the number A is congruent on division by 3 to the number

$$1^{16} - (-1)^{25}(-1)^{15} = 1 - 1 = 0;$$

that is, the desired remainder is equal to 0, and A is divisible by 3.

Example 2.2. Find the remainder on division of the same number A by 37.

For this we represent A in the following form:

$$A = (13^2)^8 - (2^5)^5 \cdot (5^3)^5 .$$

Since $13^2 = 169$ is congruent to -16, $2^5 = 32$ is congruent to -5, and $5^3 = 125$ is congruent to 14 modulo 37, the entire number A is congruent to

$$(-16)^8 - (-5)^5(14)^5 ,$$

or, equivalently, to

$$(16^2)^4 + (70)^5 .$$

But 16^2, that is, 256, is congruent to -3, and 70 is congruent to -4. This means that A is congruent to

$$(-3)^4 + (-4)^5$$

or, equivalently, to

$$81 - (2^5)^2 ,$$

and, therefore, to

$$81 - (-5)^2 = 81 - 25 = 56 .$$

Finally, 56 is congruent to 19 modulo 37. Since 19 is nonnegative and smaller than 37, it is the desired remainder.

Problem 2.1. Find the remainder on division
(a) of $A = (116 + 17^{17})^{21}$ by 8;
(b) of $A = 14^{256}$ by 17.

Problem 2.2. Prove that for any n:
(a) $6 | n^3 + 11n$;
(b) $9 | 4^n + 15n - 1$;
(c) $3^{n+2} | 10^{(3^n)} - 1$;
(d) for any a,

$$(a^2 - a + 1) | (a^{2n+1} + (a - 1)^{n+2}).$$

2.3

If a and b are congruent modulo m, we write

$$a \equiv b(\text{mod } m)$$

and say "a is congruent to b modulo m." This formula itself is called a *congruence*.

Congruence of two numbers with regard to some fixed modulus m, or equivalently, their leaving the same remainder on division by m, is also a relation that connects integers with other integers.

Some of the properties of the relation of congruence modulo m are:

Property 2.1. Reflexivity: $a \equiv a(\text{mod } m)$. Indeed, $m|a - a = 0$.

Property 2.2. Symmetry: if $a \equiv b(\text{mod } m)$, then $b \equiv a(\text{mod } m)$. Indeed, if $m|a - b$, then (by theorem 1.5) $m|b - a$.

Property 2.3. Transitivity: If $a \equiv b(\text{mod } m)$ and $b \equiv c(\text{mod } m)$, then $a \equiv c(\text{mod } m)$.

As proof of this, it is sufficient to note that from $m|a - b$ and $m|b - c$, it follows, by theorem 1.6, that $m|a - c$.

If a relation (which we shall denote by $\sim$) has the properties of reflexivity, symmetry, and transitivity, then it is called a relation of *equivalence*, or an *equivalence relation*. The simplest example of an equivalence relation on a set of numbers is the relation of equality.

Problem 2.3. Show that an equivalence relation $\sim$ on a set of numbers breaks that set up into classes (called *equivalence classes*) which satisfy the following two conditions: Any two members of the same equivalence class are connected by the equivalence relation, while no two numbers of different classes are connected by the relation.

In this problem we are concerned only with equivalence relations over the integers. This limitation is not essential, however, and the assertion of the problem is valid for equivalence relations connecting completely arbitrary objects.

Since the relation of congruence modulo m is an equivalence relation, it also breaks up, or *partitions*, the set of integers into classes. These classes are called *residue classes* modulo m.

Property 2.4. The number of residue classes modulo m is equal to m. Indeed, two numbers a and b belong to the same residue class modulo m if and only if they give the same remainder on division by m. But the remainder on division by m must be one of exactly m numbers: $0, 1, 2, \ldots,$ $m - 1$. Consequently, the number of residue classes is m.

We note one extraordinarily interesting fact.

In order for each residue class modulo m_1 to be contained in some residue class modulo m_2, it is necessary and sufficient that $m_2|m_1$.

Indeed, let us consider the residue class K_1 modulo m_1 containing the number 0. Clearly, the class K_1 consists of all numbers that give a remainder of 0 on division by m_1, that is, that are divisible by m_1. In particular, it contains the number m_1. Any residue class modulo m_2 that contains K_1 also contains 0 and therefore consists of all numbers divisible by m_2. Since m_1 is in K_1, it must be the case that $m_2 | m_1$, which proves the necessity. But the sufficiency is clear and may be easily proved.

Thus, the relation of divisibility may be defined in terms of residue classes. This method is useful because it allows us to define divisibility for objects much more general and complicated than the natural numbers. In addition, the further development of this idea leads to the theory of groups, an important branch of modern algebra that has applications to theoretical physics and crystallography.

We shall consequently continue to enumerate the properties of the congruence of numbers. From theorem 2.2 we immediately observe:

Property 2.5. If $a \equiv b(\mathrm{mod}\ m)$ and $c \equiv d(\mathrm{mod}\ m)$, then

$$a + c \equiv b + d(\mathrm{mod}\ m).$$

COROLLARY. If $a \equiv b(\mathrm{mod}\ m)$, then

$$a + r \equiv b + r(\mathrm{mod}\ m)$$

for any integer r.

Property 2.6. If $a \equiv b(\mathrm{mod}\ m)$ and $c \equiv d(\mathrm{mod}\ m)$, then

$$ac \equiv bd(\mathrm{mod}\ m).$$

Properties 2.5 and 2.6 show that, like equalities, congruences may be added and multiplied termwise.

Problem 2.4. If a given equivalence relation $\sim$ on the set of integers partitions this set into m classes and is such that from $a \sim b$ and $c \sim d$ it follows that $a + c \sim b + d$, then that relation is the relation of congruence modulo m (that is, $a \sim b$ if and only if $a \equiv b(\mathrm{mod}\ m)$).

Problem 2.5. Formulate and prove a rule for replacing any congruence by an equivalent congruence containing smaller numbers.

Problem 2.6. If the number p is prime and a is not divisible by p, then no two of the numbers $a, 2a, 3a, \ldots, (p - 1)a$ are congruent to each other modulo p. Therefore, on dividing the numbers $a, 2a, 3a, \ldots, (p - 1)a$ by p, we obtain each remainder except 0 exactly once.

Problem 2.7 (Wilson's theorem). A necessary and sufficient condition for a number p to be prime is that $(p - 1)! + 1 = 1 \cdot 2 \cdots (p - 1) + 1$ be divisible by p.

Problem 2.8. Formulate and prove the analogue of theorem 1.16 for the relation of congruence.

3 Criteria for Congruence and Criteria for Divisibility

3.1

An extremely general method for finding the remainder on division of an arbitrary, but fixed, natural number a by a given natural number m consists in the following: One constructs a sequence of natural numbers

$$a = A_0, A_1, A_2, \ldots \tag{3.1}$$

that are congruent modulo m. One must choose the sequence so that, after each of its elements that is greater than or equal to m, there is at least one more element. Then the last member of the sequence (if, of course, this exists) will clearly be the remainder r on division of a by m.

The simplest example of such a sequence is the sequence (1.3) of section 1.11. The problems on finding the remainder in examples 2.1 and 2.2 of the preceding chapter also reduce in essence to the construction of a sequence of this type.

We shall call any method for constructing a sequence of the form (3.1) a *criterion for congruence* modulo m.

In particular, one criterion for congruence modulo m is the process of successive subtraction of the number m until the first number smaller than m is obtained.

3.2

It is apparent that for a criterion for congruence to be of any use to us, it must satisfy the following three requirements:

Property 3.1. The criterion must be *well defined*; that is, the number a must completely determine all the elements of the sequence (3.1).

22

Property 3.2. The criterion for congruence must be applicable to any natural number a. This property of the criterion is called its *universality*.

Property 3.3. Finally, we must have some guarantee that in the sequence (3.1) there is at least one element that is smaller than m, and that the sequence terminates at this number. The process of constructing the sequence should, by this requirement, not continue indefinitely, but end sooner or later with the appearance of the remainder of the division of a by m. This property of a criterion for congruence is called its *determinacy*.

The requirements we have listed above can be fulfilled in a number of ways. The most natural of them is the following:

We attempt to find a function $f(x)$ subject to the following conditions:

a. $f(x)$ is a natural number for $x \geq m$;

b. $f(x)$ is not defined for $x < m$ (that is, $f(x)$ has no meaning for such an x);

(There is nothing surprising in the idea of a function losing its meaning for some values of the argument. For example, the function $1/x(x - 1)$ is undefined for $x = 0$ and $x = 1$.)

c. if $f(x)$ has meaning, $f(x) < x$;

d. if $f(x)$ has meaning, then the numbers x and $f(x)$ are congruent modulo m.

Such functions have already been shown to exist. An example is the function $f_0(x)$:

$$f_0(x) = \begin{cases} x - m & \text{if } x \geq m, \\ \text{undefined} & \text{if } x < m. \end{cases}$$

It is precisely this function that is employed in the construction of the sequence in chapter 1.

To each function $f(x)$ that satisfies conditions (a) through (d), there corresponds a unique method of constructing a sequence (3.1); that is, a unique criterion for congruence modulo m.

Indeed, given an arbitrary natural number a, let us construct the sequence of numbers

$$A_0, A_1, A_2, \ldots, \tag{3.2}$$

where $A_0 = a$, and $A_{k+1} = f(A_k)$ for $k = 0, 1, \ldots$.

If $A_k \geq m$, the number $f(A_k)$ is defined, and there is at least one element after A_k. If, however, $A_k < m$, then $f(A_k)$ is not defined, and A_k is the last element of the sequence (3.2).

Thus, we already have at least one criterion for congruence.

3.3

We shall now show that this criterion has the properties 3.1, 3.2, and 3.3.

The first requirement is fulfilled because each element of the sequence (3.2) defines the one after it uniquely (if, of course, this next element actually exists).

Here there is a certain fine point which, although not immediately apparent, is extremely important. The point is that in defining the sequence (3.2), we must first determine whether the number $f(A_k)$ exists before we can calculate its value. In other words, we must know whether the number A_k is larger or smaller than m. If the numbers A_k and m are given in a form that is convenient for comparison, say in their decimal representation, then this decision can be made without difficulty. But the comparison of the magnitudes of numbers like $2^{20} - 3 \cdot 5^2 \cdot 11 \cdot 31 \cdot 41$ and $3^{10} - 78 \cdot 757$ would require considerable effort even though the first one is 1 and the second 3.

As a result of this difficulty, we shall in the future apply criteria for congruence exclusively to positive numbers that are written in the decimal system.

As to the second requirement, it is sufficient to note the following: If $a \geq m$, we may, in fact, start the construction of the sequence once we have calculated the value of $f(a)$, which, by assumption, exists. If $a < m$, then $f(a)$ has no meaning. But, in this case, a is its own remainder on division by m; that is, the number a comprises the entire sequence (3.2).

We pass to the third requirement. By assumption, the function $f(x)$ is such that all the elements of sequence (3.2) are nonnegative, and such that the terms of this sequence decrease strictly. Since the first element of the sequence is positive, the sequence must terminate. (As can easily be verified, the index of the last element does not exceed a.) If this last element (which we shall denote by α) were greater than or equal to m then there would exist a value $f(\alpha)$, nonnegative as before, and smaller than α. This means that α would not be the least nonnegative element of our sequence, which is a contradiction. The process of constructing the sequence thus must end, and the last element of the sequence is the remainder on division of a by m.

In summary, we have established that criteria for congruence of the type that we have described meet the three requirements of being precisely defined, universal, and determinate. Processes having these three properties, which are called *algorithms*, are beginning to play a very important role in modern mathematics.[1] Some simple examples of algorithms were given at the end of section 1.1 of chapter 1. We shall become acquainted with other examples below.

3.4

One of the most important algorithms in mathematics is the so-called *Euclidean algorithm*, which consists of the following:

Let a and b be two natural numbers, with $b < a$. We divide a by b with remainder: $a = bq_0 + r_1$, where $0 \leq r_1 < b$. If $r_1 \neq 0$, then we may divide b by r_1 with remainder: $b = r_1q_1 + r_2$, with $0 \leq r_2 < r_1$. Continuing in this manner, we obtain the equations $r_1 = r_2q_2 + r_3$, $r_2 = r_3q_3 + r_4$, and so on.

We shall show that the process we have described is actually an algorithm; that is, that it has the properties of being precisely defined, universal, and determinate.

We note that the process we are considering consists of successive divisions with remainder. Therefore, the properties of being precisely defined and universal are consequences of the existence and uniqueness of the partial quotient and remainder.

It is also fairly simple to show that our process is determinate. The number b and the remainders on division form a decreasing sequence of nonnegative numbers:

$$b, r_1, r_2, \ldots . \tag{3.3}$$

But there are only $b + 1$ nonnegative numbers that are not greater than b. Sequence (3.3), therefore, cannot have more than $b + 1$ elements, so that the process cannot consist of more than b divisions with remainder.[2] Thus, the process we are considering is indeed an algorithm and fully deserves its name.

Let us clarify the circumstances under which the process ends. Clearly, the last division must be such that further division by its remainder is impossible. But this can happen only if this last remainder is equal to zero; that is, only if the last division can be performed evenly.

Problem 3.1. a. If Euclid's algorithm is applied to two numbers a and b, the last nonvanishing remainder r_n will equal (a, b).

b. For any natural numbers a and b, there exist integers A and B such that $aA + bB = (a, b)$.

Problem 3.2. From the result of part (b) of problem 3.1, derive theorems 1.9, 1.12, 1.13, and 1.14. (We emphasize that our arguments about the Euclidean algorithm were based on the possibility of division with remainder. We did not use theorems 1.9–1.14 in them, nor did we use any other considerations based on the fundamental theorem of arithmetic.)

1. For a further discussion of algorithms, see the book by B. A. Trakhtenbrot, *Algorithms and Automatic Computing Machines* (Boston: D. C. Heath, 1963).

2. In fact, the number of these divisions cannot exceed 5 log b. For a proof, see pp. 34–35 of the author's book, *The Fibonacci Numbers*, in the Topics in Mathematics series (D. C. Heath and Co., Boston: 1963).

3.5

Of course, the description of an algorithm given in section 3.3 is not the precise definition of this notion, which is comparatively complicated and cannot be formulated here.[3] The requirements listed do, however, reflect fairly completely the conditions that algorithms must satisfy. Algorithms are important because they provide a uniform method for solving all problems of a given type. For example, the algorithms which we have just considered allow us to calculate the remainder on division of any number a by some fixed number m.

Any method of computing using formulas in which numbers are substituted for variables is also an algorithm. Speaking somewhat more freely, we can say that all procedures which can be carried out by machines are algorithms. It is no accident, therefore, that the development of the theory of algorithms coincided historically with the appearance and widespread use of computers; but computational problems in the narrow sense of the word—that is, problems for which a numerical answer may be obtained from initial data by some type of mechanical rule—are not the only problems that reduce to algorithms. One may, for example, attempt to find an algorithm that will provide the proof of any true proposition in some branch of mathematics. Such an algorithm must be able to turn the formulation of theorems into their proofs. Fantastic as this may seem, such algorithms do in fact exist although not for very wide areas of mathematics. At the same time, however, the existence of such algorithms may be proved impossible in some branches of mathematics (such as all branches containing all of arithmetic).

3.6

Let us now find and examine several criteria for congruence using the method presented in section 3.1. Here and in the future we shall assume that all numbers are written in decimal notation.

First we shall find a criterion for congruence modulo 5.

Let A be a natural number. We shall represent A in the form $10a + b$ (b is just the last digit in the decimal representation of the number A), and set

$$f_1(A) = \begin{cases} b \text{ if } A \geq 10\,, \\ b - 5 \text{ if } 5 \leq A < 10\,, \\ \text{not defined if } A < 5\,. \end{cases}$$

3. For this definition, see the book by Trakhtenbrot referred to in footnote 1 of this chapter.

The reader may check for himself that the function f_1 satisfies conditions (a) to (d) of section 3.2.

Thus, to find the remainder on division of some number by 5, it is sufficient to consider the last digit. If this digit is smaller than 5, then it is the desired remainder. If not, subtracting 5 from it gives the remainder. We note that the application of this criterion for congruence to any number reduces to the construction of a sequence of the type (3.2) that consists of not more than three elements.

Of course, the object of all these arguments is not to discover this well-known criterion for divisibility by 5, but to express it in the general form that was described in section 3.2.

Problem 3.3. State and analyze analogous criteria for congruence modulo 2, 4, 8, 10, 16, 20, and 25.

Problem 3.4. Suppose some natural number $m \neq 0$ is fixed. Let us represent each natural number A in the form

$$A = 10^k a + b \qquad (0 \leq b < 10^k),$$

and set

$$f(A) = \begin{cases} b \text{ if } A \geq 10^k, \\ \text{the remainder on division of } A \text{ by } m \text{ if } m \leq A < 10^k, \\ \text{undefined if } A < m. \end{cases}$$

For what numbers m will such an algorithm be a criterion for congruence modulo m for some k?

3.7

As a second example, we shall consider a criterion for congruence modulo 3.

Let us first represent each natural number A in the form

$$A = 10^n a_n + 10^{n-1} a_{n-1} + \cdots + 10 a_1 + a_0,$$

where $0 \leq a_i < 10$ (the numbers $a_0, a_1, \ldots, a_{n-1}, a_n$ are the decimal digits of A). Let us then define

$$f_2(A) = \begin{cases} a_0 + a_1 + \cdots + a_{n-1} + a_n, \text{ if } A \geq 10, \\ \text{the remainder on division of } A \text{ by 3 if } 3 \leq A < 10, \\ \text{undefined if } A < 3. \end{cases}$$

Problem 3.5. Verify that the function $f_2(x)$ satisfies conditions (a) through (d) of section 3.2 and thus defines a criterion for congruence modulo 3.

Problem 3.6. Apply the above criterion for congruence modulo 3.
a. to the numbers 858,733 and 789,988;
b. to the number whose decimal representation consists of 4444 fours.

Problem 3.7. State and analyze analogous criteria for congruence modulo 7, 9, 11, 13, and 37.

3.8

In many cases, it is not only the magnitude of the partial quotient of two numbers that is nonessential. Often the magnitude of the remainder is unimportant and all that matters is whether or not it is zero; that is, whether the first number is divisible by the second. Using the methods of section 3.1, it is clear how we may approach such problems.

We shall call the numbers a and b *equidivisible* by m if either both a and b are divisible by m or if neither a nor b is divisible by m.

Problem 3.8. Prove that for any number m, all numbers congruent modulo m are equidivisible by m. Show, by example, that the converse is false.

Problem 3.9. For what numbers m does equidivisibility of two numbers by m imply congruence modulo m?

Problem 3.10. Prove that the relation of equidivisibility by a given number m is an equivalence relation, and that it partitions the set of integers into two classes.

Problem 3.11. Does theorem 2.2 hold for equidivisible numbers? Does its corollary?

3.9

Suppose that it is necessary to determine the divisibility of the number A by m. We construct a sequence of integers with decreasing absolute value,

$$A = A_0, A_1, A_2, \ldots, \tag{3.4}$$

that are equidivisible with A on division by m with remainder. The sequence (3.4) is constructed so that, after every element that is greater than or equal to m in absolute value, there follows at least one element. If the last element of this sequence is zero, then A is divisible by m, and if not, then it is not divisible by m.

We shall call any method for constructing such a sequence (3.4) a *criterion for divisibility* by m.

Problem 3.12. Prove that any criterion for congruence modulo m is a criterion for divisibility by m.

Obviously, criteria for divisibility must satisfy the same requirements of precise definition, universality, and determinacy that we require of criteria for congruence.

It is not difficult to verify (and this is left to the reader) that, by using any function $f(x)$ that satisfies conditions (a) through (c) of section 3.2 and condition (d*)—if $f(x)$ has meaning, then the numbers x and $f(x)$ are equidivisible on division by m—it is possible to construct a criterion for divisibility by m in the same manner that we constructed a criterion for congruence modulo m using any function satisfying conditions (a) through (d).

Let us construct some criteria for divisibility. Because of theorem 1.16, it is sufficient to construct only criteria for divisibility by numbers of the form p^{α} (that is, by powers of primes).

3.10

A criterion for divisibility by 7. Let A be a natural number. We represent A in the form $10a + b$, where $0 \le b < 10$. Let us set

$$f_3(A) = \begin{cases} a - 2b \text{ if } A \ge 190, \\ \text{the remainder on division by 7 for } 7 \le A < 190, \\ \text{undefined for } A < 7. \end{cases}$$

Problem 3.13. Show that the function $f_3(A)$ satisfies conditions (a) through (c) and (d*).

The function $f_3(A)$ gives us a well-known criterion for divisibility by 7: The number $10a + b$ $(0 \le b < 10)$ is divisible by 7 if and only if the number $a - 2b$ is divisible by 7. The number so obtained is checked for divisibility by 7 by the same method, and so on.

Problem 3.14. Prove that the criterion for divisibility by 7 that we have just constructed is not a criterion for congruence modulo 7.

3.11

A criterion for divisibility by 13. Let us represent each natural number A in the form $10a + b$ and set

$$f_4(A) = \begin{cases} a + 4b \text{ if } A \geq 40, \\ \text{the remainder on division of } A \text{ by 13 if } 13 \leq A < 40, \\ \text{not defined if } A < 13. \end{cases}$$

Problem 3.15. Verify that the function $f_4(x)$ satisfies conditions (a) to (c) and (d*), and formulate the criterion for divisibility by 13 so obtained.

Problem 3.16. What would be the consequences of changing the number 40 to a smaller one in the definition of the function f_4?

Problem 3.17. Construct criteria for divisibility by 17, 19, 23, 29, and 31 by analogy with the criteria for 7 and 13.

Problem 3.18. Construct two criteria for divisibility by 49.

3.12

In the preceding sections of this chapter, we became acquainted with a large number of widely varying criteria for congruence and criteria for divisibility. The goal in constructing all of these criteria is to obtain practical algorithms for finding the remainders on division by certain numbers (criteria for congruence) or for deciding whether these remainders are equal to zero (criteria for divisibility). To what extent have we carried out our proposed task?

Some criteria for congruence, such as those for division by 2, 3, 4, and 10, actually turned out to be extremely practical and useful. Others involved clumsy calculations and were therefore less practical.

It is natural, therefore, to seek and use the most convenient and efficient possible criteria for divisibility and congruence.

One of the difficulties we encounter in such attempts is that we must know how simple (or how complicated) it would be to use a given criterion in the case of some particular number. We may, for example, take as a measure of efficiency the number of arithmetic operations on single digit numbers that must be performed in using a given criterion on one number or another.

Unfortunately, any such numerical characterization of the amount of calculation to be done would depend very much on the individual properties of the number whose divisibility we want to test.

Thus, it is very easy to decide that the remainder on division of 31,025 by 8 is 1. To do this it is sufficient to find the remainder on division of 25 by 8. But to find the remainder on division of 30,525 by 8, it is necessary to divide 525 by 8 with remainder, and this requires a larger number of calculations (regardless of whether they are done mentally or on paper).

As another example, we may consider the criterion for divisibility by 37 (see problem 3.7). The remainder on dividing 10,014,023 by 37 can be found by the addition $10 + 14 + 23$ and the subsequent division by 37 of the sum obtained. As can easily be seen, this is equal to 10. However, there are very few people who could apply this criterion mentally to the number 782,639,485.

In speaking about the convenience of using various criteria for divisibility and congruence, we must therefore look beyond the difficulties of applying these tests to particular numbers and evaluate the effectiveness of each criterion "on the average." Only with such an approach can we hope to formulate precisely the degree of intricacy of a criterion for divisibility or congruence, or to find the criterion that is most economical in the majority of cases. Unfortunately, these questions are very complicated and we will not be able to develop them here.

3.13

All of the criteria for congruence and divisibility that we have constructed above appear to be a bit artificial, and at first glance it may seem that at least some of these criteria were discovered by accident or as the result of trial and error. In fact, this is not so. Indeed, it may be shown that there exist methods of constructing criteria for divisibility and congruence for any previously given number. These methods are called *general criteria for divisibility* and *general criteria for congruence*, respectively.

General criteria for divisibility are thus *methods* for obtaining actual criteria for divisibility. Actual criteria, therefore, may be considered to be the results obtained by applying general criteria to particular numbers. From this standpoint, the general criteria for divisibility are related to the actual ones in the same way that an actual criterion is related to the result of its application to a specific number, that is, to the remainder on division of a given number a by a given number m.

General criteria for divisibility and for congruence resemble algorithms; in fact, they seem to be special algorithms whose end products are themselves algorithms—namely, actual criteria for divisibility or congruence.

But before we may talk about general criteria for divisibility and congruence as algorithms, we must convince ourselves that they have the necessary properties of precise definition, universality, and determinacy.

Speaking in even more detail, we must verify that the following conditions are satisfied by any general criterion for divisibility or for congruence. First, given any number m, the general criterion must actually yield a criterion for divisibility (or for congruence) for m. It must, so to speak, "convert" each natural number m into the corresponding criterion. This is merely a precise statement of the property of *determinacy*. Second, a general criterion must be *precisely defined*, that is, when applied to any given number m it must, in a definite manner, lead to a well-defined criterion for divisibility (or for congruence) for this number. Third and last, a general criterion should be *universal*, that is, truly general, and should give a criterion for divisibility or congruence for any natural number given in advance.

In this sense, neither the method described in section 3.2 for obtaining a criterion for congruence nor the method described in section 3.6 of finding criteria for divisibility are general criteria. In fact, our procedure for finding functions to satisfy the necessary conditions satisfies none of the conditions of precise definition, universality, or determinacy.

Indeed, these methods give no guarantee that the necessary function will be found; hence, they are not determinate. Furthermore, if the necessary function does exist, it can be found in various ways, to say nothing of the fact that there may be several functions satisfying our conditions. This means that the result of our methods is not precisely defined. Finally, our techniques are not universal enough, since it is quite possible that we shall be unable to find the required functions for some numbers. In any event, the method itself gives us no specific guarantee of success. Thus, for the described process to become an algorithm, it will have to be supplemented with some precise instructions that will guarantee the construction of a completely determined function f_m for each number m.

This problem of the "algorithmization" of the construction of criteria for divisibility can, in fact, be solved without particular difficulty, and general criteria for divisibility have been known for several centuries.

We have actually constructed one such general criterion for congruence already in our treatment in Chapter 1, Section 11, about division with remainder. We may formulate it as follows: To each positive integer m there corresponds the process of successive subtraction of m from any other number k until a number is obtained that is smaller than m (see the last sentence of section 3.1). It is clear that such a correspondence has the necessary properties of precise definition (we know exactly what process corresponds to the number m), universality (the process of successive subtraction can be carried out with any m), and determinacy (such an attempt inevitably leads to success). But the practical value of the general criterion for congruence just described is very small.

A certain improvement of the general criterion for congruence based on successive subtractions leads to the familiar process of long division, which is itself a general criterion for congruence. It is worth noting that most people use precisely this criterion for finding remainders on division. The reasoning may go according to the scheme outlined in Table 3.1, in which we give two versions—one in the usual, everyday language, and one in the language of algorithms.

Table 3.1

In everyday language	In the language of algorithms
1. I must find the remainder on division of a given a by a given m;	The general criterion for congruence begins to process the number m;
2. to do this I must divide by m;	the general criterion "issues" the result of its processing of the number m: an actual criterion for congruence on division by m, which consists in dividing by m directly;
3. now I begin to divide a by m . . .	the actual criterion obtained begins to process the number a;
4. I divide and obtain the remainder.	the actual criterion reaches its goal: the remainder on division of a by m.

Since the first three steps of this reasoning are very simple, we should not be surprised that the fourth step—the actual performance of division—turns out to be so unwieldy. The goal in creating general methods for congruence and divisibility consists precisely in lightening the load of the fourth step at the expense of increasing the difficulty of the second. It is exactly this that we have in mind when we speak of general criteria for divisibility and congruence.

3.14

Historically, the first useful general criterion for divisibility (in fact, it is even a general criterion for congruence) is the following method, proposed early in the seventeenth century by the famous French

mathematician and philosopher Blaise Pascal. Its essence is as follows:
Let m be a natural number. We form the sequence of numbers

$$r_1, r_2, r_3, \ldots , \qquad\qquad (3.5)$$

by setting

r_1 equal to the remainder on dividing 10 by m ,
r_2 equal to the remainder on dividing $10r_1$ by m ,
r_3 equal to the remainder on dividing $10r_2$ by m ,

and so on.
Let us now represent an arbitrary number A in the form

$$10^n a_n + 10^{n-1}a_{n-1} + \cdots + 10a_1 + a_0 ,$$

(where $0 \le a_i < 10$ for all i), and define the function

$$F_m(A) = \begin{cases} a_0 + r_1a_1 + r_2a_2 + \cdots + r_na_n \text{ if } 10^n \ge m , \\ \text{the remainder on division of } A \text{ by } m \text{ if } 10^n < m \le A , \\ \text{undefined if } A < m . \end{cases}$$

Problem 3.19. Verify that for any m the function F_m satisfies conditions (a) through (d) of section 3.2.

Thus, we have developed a means for constructing a criterion for congruence modulo any m; that is, a general criterion for congruence.

Problem 3.20. Formulate the actual criteria for congruence
a. modulo 2, 5, and 10;
b. modulo 4, 20, and 25;
c. modulo 3 and 9;
d. modulo 11;
e. modulo 7;
which are obtained from Pascal's general criterion for congruence.

Problem 3.21. Suppose that in the sequence (3.5), r_1 is the remainder on division of 100 by m, r_2 is the remainder on division of $100r_1$ by m, r_3 is the remainder on division of $100r_2$ by m, and so on. Use these values of the r_i to derive a general criterion for congruence that is analogous to that of Pascal.

3.15

In section 3.12 we spoke of the comparative advantages of different criteria for divisibility (or congruence) for a given number. Since a general criterion for divisibility must give us criteria for divisibility by any natural number, it is not surprising that the criteria it gives us for different numbers may differ greatly in efficiency.

For example, Pascal's general criterion for divisibility gives wholly acceptable criteria for congruence modulo 3 and 11, together with a very clumsy and inconvenient criterion for congruence modulo 7 (see problem 3.20, part e).

In connection with this, we can make observations about general criteria for divisibility and congruence that are similar to those made in our discussion in section 3.12 on the practicality of actual criteria for divisibility. In this sense the best possible general criterion for divisibility (or congruence) should be the one which, on application to any previously given positive integer m, gives the best possible criterion for division (congruence) for this m. The problem of finding the best general criterion for divisibility is not only far from being solved, but is also far from being precisely formulated.

4   Divisibility of Powers

4.1

The question of the divisibility of powers is, in reality, a question about the divisibility of a certain type of product, namely, the product of several equal factors. It may be answered, therefore, on the basis of the results of chapter 2. In the case of large exponents, however, decreasing the base of a power may not lead immediately to the remainder on division of the power by some given number, and we must use some artificial means to obtain this remainder (see the examples in section 2.2). Moreover, in constructing general criteria for divisibility, we required calculation of the remainder on division of successive powers of 10^k by m. Although this process in itself is not complicated, it still does not indicate any regularities in sequence (3.5), nor does it make it possible to choose the number k so that all of these remainders are small. At the same time, such a possibility exists; in fact, the number k may be chosen so that all of these remainders are 1.

These considerations make it desirable to study the divisibility of powers in more detail.

4.2

Let us begin by establishing several number-theoretic results.

THEOREM 4.1 (Fermat's little theorem). *If the number p is a prime, then for any positive integer a the difference $a^p - a$ is divisible by p.*

One should not confuse "Fermat's little theorem" with "Fermat's last theorem." The latter asserts that for any integer $n > 2$ there do not exist integers a, b, c satisfying $a^n + b^n = c^n$. Despite numerous attempts, Fermat's last theorem has yet to be either proved or refuted.

COROLLARY. *If p is a prime and a is not divisible by p, then $a^{p-1} - 1$ is divisible by p.*

Problem 4.1. Give an example showing that both theorem 4.1 and its corollary are, in general, not true if p is a composite number.

Problem 4.2. Prove Fermat's little theorem using the result of problem 2.6.

Suppose that a natural number $m > 1$ has the canonical decomposition

$$m = p_1{}^{\alpha_1} p_2{}^{\alpha_2} \cdots p_k{}^{\alpha_k} . \tag{4.1}$$

We set

$$\varphi(m) = p_1{}^{\alpha_1-1}(p_1 - 1) p_2{}^{\alpha_2-1}(p_2 - 1) \cdots p_k{}^{\alpha_k-1}(p_k - 1) \tag{4.2}$$

and define $\varphi(1) = 1$.

Formulas (4.1) and (4.2) assign to each positive integer m a unique number $\varphi(m)$. This means that we may speak of φ as a function on the positive integers.

DEFINITION 4.1. *The function φ above is called* Euler's function.

Euler's function plays an exceptionally important role in many number-theoretic questions. Several of these applications will be discussed later in this book.

THEOREM 4.2. *If m_1 and m_2 are relatively prime, then*

$$\varphi(m_1 m_2) = \varphi(m_1)\varphi(m_2) .$$

Problem 4.3. Calculate $\varphi(12)$, $\varphi(120)$, $\varphi(1000)$.

Problem 4.4. Determine all numbers m for which
a. $\varphi(m) = 10$;
b. $\varphi(m) = 8$.

Problem 4.5. Prove that there exists no number m for which $\varphi(m) = 14$.

Problem 4.6. Show that $\varphi(m)$ is equal to the number of natural numbers relatively prime to m which are less than or equal to m. This property of Euler's function is extraordinarily important. It is often taken as the definition of this function.

THEOREM 4.3 (Euler's theorem). *If the numbers a and m are relatively prime, then $a^{\varphi(m)} - 1$ is divisible by m.*

4.3

We shall now construct several general criteria for divisibility and for congruence, using the theorems stated above.

Let us fix the natural number m and represent the natural number A in the form

$$A = a_0 + a_1 10^{\varphi(m)} + a_2 10^{2\varphi(m)} + \cdots + a_k 10^{k\varphi(m)},$$

where

$$0 \le a_i < 10^{\varphi(m)} \text{ for all } i \text{ from 1 to } k;$$

that is, the numbers a_i ($i = 0, 1, \ldots, k$) have at most $\varphi(m)$ digits.

The function F defined by

$$F(A) = \begin{cases} a_0 + a_1 + \cdots + a_k \text{ if } A \ge 10^{\varphi(m)}; \\ \text{the remainder on division of } A \text{ by } m \text{ if } m \le A < 10^{\varphi(m)}; \\ \text{undefined if } A < m, \end{cases}$$

may easily be shown to determine a general criterion for congruence modulo any m.

Problem 4.7. Verify this.

THEOREM 4.4. *If the numbers a and m are relatively prime and the numbers k_1 and k_2 are congruent modulo $\varphi(m)$, then the numbers a^{k_1} and a^{k_2} are congruent modulo m.*

Problem 4.8. Prove that $2730 | n^{13} - n$ for any n.

4.4

This general criterion for congruence is often not "sufficiently economical," since the number $\varphi(m)$ frequently turns out to be fairly large. In applying this criterion, it is often necessary to add large numbers and to divide $\varphi(m)$-digit numbers directly by m (or to make use of some other criterion for divisibility or congruence). It would thus be desirable to replace $\varphi(m)$ by a smaller exponent, and this can in fact be done in many cases. For example, for $m = 37$, instead of $\varphi(m) = 36$ we may take the exponent 3, since $10^3 = 1000$ leaves a remainder of 1 on division by 37, as must 10^{36} by Euler's theorem; for $m = 11$, instead of $\varphi(m) = 10$ we may take the exponent 2, and so on.

DEFINITION 4.2. *The least number δ for which a^δ leaves a remainder of 1 on division by m is called the* order *of a on division by m with remainder.*

This number is more often called the *order of a modulo m*.

Euler's theorem makes it clear that for any relatively prime numbers a and m, the order of a modulo m does not exceed $\varphi(m)$. This order may be taken in place of $\varphi(m)$ in the formulation of the general criterion given in section 4.3.

Problem 4.9. Modify the general criterion for congruence constructed above by replacing $\varphi(m)$ with the order of 10 modulo m, and verify that the modified criterion is itself a general criterion for congruence.

4.5

Applications of Euler's function and Euler's theorem are not limited to criteria for divisibility. Another important use lies in the solution of equations in integers, often called Diophantine equations.

THEOREM 4.5. *If the numbers a and b are relatively prime, then the equation*

$$ax + by = c \qquad (4.3)$$

is always solvable in integers, and all its solutions in integers are pairs of the form (x_t, y_t), where

$$x_t = ca^{\varphi(b)-1} + bt$$
$$y_t = c\frac{1 - a^{\varphi(b)}}{b} - at,$$

for any integer t.

Problem 4.10. Prove a theorem analogous to theorem 4.5 without the hypothesis that a and b are relatively prime.

Problem 4.11. Find a method for solving equation (4.3) in integers on the basis of problem 3.1.

Problem 4.12. Solve the following equations in integers:
a. $5x + 7y = 9$;
b. $25x + 13y = 8$.

Chapter Four

4.6

THEOREM 4.6. *Let m be relatively prime to* 10, *and let k be congruent to* $10^{\varphi(m)-1}$ *modulo m. Then the numbers* $10a + b$ *and* $a + kb$ *are equidivisible by m.*

We can use this theorem to construct the following general criterion for divisibility: Let k' denote the remainder on division of $10^{\varphi(m)-1}$ by m. We represent an arbitrary number A in the form $10a + b$ $(0 \le b < 10)$ and set

$$F(A) = \begin{cases} a + k'b, \text{ for } A > a + k'b; \\ \text{the remainder on division of } A \text{ by } m, \text{ for } m \le A \le a + k'b; \\ \text{undefined for } A < m. \end{cases}$$

If k' is large (close to m), we may conveniently replace it by the negative number $k' - m$ in the formulation of the corresponding criterion.

Problem 4.13. Verify that conditions (a) through (c) and (d*) are satisfied for the function F.

Problem 4.14. Use the general criterion just constructed to derive criteria for divisibility by the numbers 17, 19, 27, 31, and 49.

Problem 4.15. Construct an analogous general criterion for divisibility, using the fact that one may represent any arbitrary natural number in the form $100a + b$ $(0 \le b < 100)$, and from it derive criteria for divisibility by 17, 43, 49, 67, 101, and 109.

Proofs of
Theorems

THEOREM 1.1. It is sufficient to note that $a = a \cdot 1$.

THEOREM 1.2. By hypothesis, there exist d_1 and d_2 such that $b = ad_1$ and $c = bd_2$. But then $c = ad_1d_2$, and $a|c$.

THEOREM 1.3. We have $a = bc_1$ and $b = ac_2$, from which it follows that $a = ac_1c_2$, that is, that $c_1c_2 = 1$. Since the numbers c_1 and c_2 are integers, either $c_1 = c_2 = 1$ or $c_1 = c_2 = -1$. In the first case $a = b$, and in the second $a = -b$.

THEOREM 1.4. Let $a = bc$. If $|c| \geq 1$, then $|b| > |a|$ implies $|bc| > |a|$, or $|a| > |a|$, a contradiction. Hence, $|c| < 1$, but since c must be an integer, $c = 0$, so that $a = bc$ is also 0.

THEOREM 1.5. Clearly, from $a = bc$ it follows that $|a| = |b||c|$, and, conversely, if $|a| = |b||c|$, then $a = b(\pm c)$. Since c is an integer if and only if $|c|$ is an integer, $b|a$ if and only if $|b| \mid |a|$.

THEOREM 1.6. We may write

$$a_1 = bc_1,$$
$$a_2 = bc_2,$$
$$\dots\dots\dots$$
$$a_n = bc_n,$$

where all of the numbers $c_1, c_2, \dots, c_n$ are integers. Adding these equations term by term, we obtain

$$a_1 + a_2 + \cdots + a_n = b(c_1 + c_2 + \cdots + c_n).$$

Since the expression inside the parentheses is an integer, the proof is complete.

THEOREM 1.8. The proof is by contradiction. Suppose that there are only a finite number of prime numbers, so that they may all be listed:

$$p_1, p_2, \dots, p_n. \tag{5.1}$$

Let us denote the product of these numbers by P and consider the sum $P + 1$. This sum is greater than every prime number given in the listing (5.1) and therefore cannot be one of these prime numbers. Consequently, it is divisible by at least one prime p_k. But P is also divisible by p_k. Thus, on the basis of the corollary to theorem 1.6, we must have $p_k | 1$, from which it follows that $p_k = \pm 1$—contradicting the definition of a prime number (see definition 1.5 in section 1.12). This proof of the existence of infinitely many primes was given by Euclid in the fourth century B.C.

THEOREM 1.9. Suppose the numbers a and p are not relatively prime. Then both are divisible by some positive integer other than 1. Since p is a prime, the only such number can be p. Thus, in this case, $p | a$. So either a and p are relatively prime, or $p | a$.

THEOREM 1.10. Dividing M by m with remainder, we obtain

$$M = mq + r,$$

where $0 \le r < m$. Since M and m are both divisible by both a and b, we know by the corollary to theorem 1.6 that the number r must also be divisible by a and by b, and thus must be a common multiple of these two numbers. But $r < m$, and m is the least positive common multiple of a and b. This means that r cannot be a positive number, so that $r = 0$. Therefore, $m | M$.

THEOREM 1.11. Suppose the numbers a and b are relatively prime, and let m be their least common multiple. Since $a | ab$ and $b | ab$, we may conclude by the preceding theorem that $m | ab$. Let $ab = mk$, and set $m = ac$. Then $ab = ack$; that is, $b = ck$, so that $k | b$. In exactly the same way, we see that $k | a$. Since the numbers a and b are relatively prime, we must have $k = 1$, which means that $m = ab$.

THEOREM 1.12. Let us denote the least common multiple of the numbers b and c by m. By the preceding theorem, $m = bc$. Furthermore, $c | ab$ by hypothesis, and it is clear that $b | ab$. It follows from theorem 1.10 that $bc | ab$; that is, that for some number k, $ab = bck$, which implies at once that $a = ck$. But this implies that $c | a$, the desired result.

THEOREM 1.13. The proof is by induction on the number of factors. If there is only one factor, the theorem is trivial. Let us suppose that the theorem is true for any product of n factors. Suppose $p | a_1 a_2 \cdots a_n a_{n+1}$. Then, denoting $a_1 a_2 \cdots a_n$ by A, we have $p | A a_{n+1}$. If $p | a_{n+1}$, then the theorem is proved. If not, then by theorem 1.9, a_{n+1} and p are relatively prime. But then, by the preceding theorem, $p | A$. Since A is a product of only n factors, the inductive hypothesis implies that one of these factors must be divisible by p, which proves the theorem.

COROLLARY. The entire fraction is an integer (that is, the numerator is divisible by the denominator) since, by the binomial theorem, it is one of the coefficients of the expansion of $(x + 1)^p$. We shall therefore consider the numerator as a product of two factors: p and $1 \cdot 2 \cdots (p - 1) = (p - 1)!$

None of the factors in the denominator is divisible by p. Hence, by the preceding theorem, the entire denominator is not divisible by p, and is therefore, by theorem 1.9, relatively prime to p. Thus the second factor of the numerator, $(p - 1)!$, must be divisible by the denominator. Denoting the quotient of this division by q, we have $\binom{p}{k} = pq$, and the proof is complete.

THEOREM 1.14. First we shall show the *possibility* of decomposing any number other than 1 into prime factors. Let us suppose that all numbers smaller than N can be so decomposed. If the number N is prime, then it may automatically be decomposed into a product of primes (namely, a product consisting of only one factor—the number N itself), and the theorem is proved. Suppose now that N is composite and that N_1 is some divisor of N that is distinct from both N and 1. Let N_2 be the quotient on division of N by N_1. Then $N = N_1 N_2$, and it can easily be verified that $1 < N_2 < N$. Since N_1 and N_2 are both smaller than N, they are both decomposable into products of prime factors (by the induction hypothesis). Let $N_1 = p_1 p_2 \cdots p_k$ and $N_2 = q_1 q_2 \cdots q_l$ be these decompositions. Then $p_1 p_2 \cdots p_k q_1 q_2 \cdots q_l$ is the desired decomposition of the number N, and the possibility of decomposition is proved.

We now prove the *uniqueness* of this decomposition. Suppose we are given two decompositions of the number N into prime factors: $N = p_1 p_2 \cdots p_k$ and $N = q_1 q_2 \cdots q_l$ (where not all of the p_i nor all of the q_j are necessarily distinct). Clearly,

$$p_1 p_2 \cdots p_k = q_1 q_2 \cdots q_l. \tag{5.2}$$

Since $q_1 q_2 \cdots q_l$ is divisible by p_1, we have by the preceding theorem that at least one of the numbers $q_1, q_2, \ldots, q_l$ is divisible by p_1. Suppose that $p_1 | q_1$ (we may make this assumption since we have the right to change the order of the factors so that q_1 is the factor which is divisible by p_1). Since the number q_1 is prime, this is possible only if $p_1 = q_1$. Dividing both sides of equality (5.2) by p_1, we obtain

$$p_2 p_3 \cdots p_k = q_2 q_3 \cdots q_l. \tag{5.3}$$

Similarly, we can see that one of the numbers $q_2, q_3, \ldots, q_l$ (say q_2) is divisible by p_2, and thus that $p_2 = q_2$. Dividing both sides of equality (5.3) by p_2, we reduce by one more the number of factors on each side. This process of reduction can clearly be carried out until one of the products has been completely cancelled. Suppose that the left side of (5.2) is the first to be cancelled completely. The product on the right side of (5.2) must then also be cancelled completely, for otherwise we would obtain an equation of the form

$$1 = q_{k+1} \cdots q_l,$$

which is impossible since 1 is not divisible by any prime. Therefore, we have shown that $k = l$, and that if the q_i are suitably reordered,

$$p_1 = q_1, p_2 = q_2, \ldots, p_k = q_k,$$

so that the two decompositions are identical up to the order of the factors. The theorem is thus proved in full.

THEOREM 1.15. Suppose that $p_1^{\alpha_1} p_2^{\alpha_2} \cdots p_k^{\alpha_k}$ and $q_1^{\beta_1} q_2^{\beta_2} \cdots q_l^{\beta_l}$ are the canonical decompositions of a and b respectively, and that d is some common divisor of a and b. If $d \neq 1$, then d is divisible by some prime p. Thus, by theorem 1.2, $p|a$ and $p|b$ so that p is among the numbers $p_1, p_2, \ldots, p_k$ as well as among the numbers $q_1, q_2, \ldots, q_l$ by theorem 1.13. Therefore, among the prime factors appearing in the canonical decomposition of a, there is at least one that appears in the canonical decomposition of b.

Conversely, if a and b are relatively prime and p appears in the canonical decomposition for a, then b is not divisible by p, so that p cannot enter into the canonical decomposition for b.

THEOREM 1.16. Suppose $a|b$. Since $p_i^{\alpha_i}|a$ $(i = 1, 2, \ldots, k)$, we obtain the desired conclusion from $a|b$ by a simple reference to theorem 1.2.

The converse is proved by induction. We assume $p_i^{\alpha_i}|b$, all i, and we suppose that we have already proved $p_1^{\alpha_1} \cdots p_l^{\alpha_l}|b$ $(1 \leq l \leq k)$. Then, since $p_{l+1}^{\alpha_{l+1}}|b$ and since $(p_{l+1}^{\alpha_{l+1}}, p_1^{\alpha_1} \cdots p_l^{\alpha_l}) = 1$, we may apply the preceding theorem and the corollary to theorem 1.11 to obtain $p_1^{\alpha_1} \cdots p_l^{\alpha_l} p_{l+1}^{\alpha_{l+1}}|b$. This completes the inductive step and the proof.

THEOREM 2.1. Suppose a and b are congruent modulo m. We write

$$a = mq_1 + r_1 \qquad (0 \leq r_1 < m), \tag{5.4}$$

$$b = mq_2 + r_2 \qquad (0 \leq r_2 < m). \tag{5.5}$$

Because of the congruence of a and b, we have $r_1 = r_2$. This means that

$$a - b = m(q_1 - q_2);$$

that is, $m|a - b$.

Conversely, suppose that $m|a - b$. Dividing a and b by m with remainder, we obtain (5.4) and (5.5). Here

$$a - b = m(q_1 - q_2) + r_1 - r_2;$$

that is,

$$(a - b) - m(q_1 - q_2) = r_1 - r_2.$$

By theorem 2.6, $m|r_1 - r_2$. Since $|r_1 - r_2| < m$, however, it follows from theorem 1.4 that $r_1 - r_2 = 0$, or $r_1 = r_2$, the required result.

THEOREM 2.2. By hypothesis and theorem 2.1 we may write

$$\left.\begin{aligned} a_1 &= b_1 + mq_1, \\ a_2 &= b_2 + mq_2, \\ &\cdots\cdots\cdots \\ a_n &= b_n + mq_n. \end{aligned}\right\} \tag{5.6}$$

Adding these equations term by term, we obtain
$$(a_1 + a_2 + \cdots + a_n) - (b_1 + b_2 + \cdots + b_n) = m(q_1 + q_2 + \cdots + q_n),$$
which, by virtue of theorem 2.1, shows the congruence of the two sums.

To prove the congruence of the two products, we make use of the following identity:

$$(d_1 + s_1 m)(d_2 + s_2 m) = d_1 d_2 + (d_1 s_2 + d_2 s_1 + s_1 s_2 m)m.$$

Arguing inductively, we can easily see that this implies

$$(d_1 + s_1 m)(d_2 + s_2 m)\cdots(d_k + s_k m) = d_1 d_2 \cdots d_k + tm,$$

where t is some integer.

If we now multiply all the equalities (5.6) term by term, and apply the arguments just given to the right-hand side, we obtain

$$a_1 a_2 \cdots a_n = b_1 b_2 \cdots b_n + mt,$$

where t is some integer. The congruence of the products modulo m is thus proved.

THEOREM 4.1. The proof is by induction on a. For $a = 1$,

$$a^p - a = 1 - 1 = 0 ,$$

and $p|0$.

Let us suppose that $a^p - a$ is divisible by p. We shall show that $(a + 1)^p - (a + 1)$ is also divisible by p. Indeed, expanding $(a + 1)^p$ by the binomial theorem, we have

$$(a + 1)^p - (a + 1) = a^p + \binom{p}{1}a^{p-1} + \binom{p}{2}a^{p-2} + \cdots$$

$$+ \binom{p}{p-1}a + 1 - a - 1$$

$$= a^p - a + \binom{p}{1}a^{p-1} + \binom{p}{2}a^{p-2} + \cdots$$

$$+ \binom{p}{p-1}a . \tag{5.7}$$

$a^p - a$ is divisible by p by hypothesis. By the corollary to theorem 1.13, $\binom{p}{k}$ is also divisible by p for $1 \le k \le p - 1$. Hence, each term of the sum (5.7) is divisible by p and therefore (by theorem 1.6), so is the entire sum.

This justifies the inductive transition, and thus the entire theorem is proved.

COROLLARY. By Fermat's theorem,

$$p|a^p - a = a(a^{p-1} - 1) .$$

If a is not divisible by p here, then by theorem 1.13, p must divide $a^{p-1} - 1$.

THEOREM 4.2. Let $m_1 = p_1^{\alpha_1} \cdots p_k^{\alpha_k}$ and let $m_2 = q_1^{\beta_1} \cdots q_l^{\beta_l}$. By theorem 1.15, each of the numbers $p_1, \ldots, p_k$ is distinct from each of the numbers $q_1, \ldots, q_l$. This means that the canonical decomposition of $m_1 m_2$ is $p_1^{\alpha_1} \cdots p_k^{\alpha_k} q_1^{\beta_1} \cdots q_l^{\beta_l}$. Therefore,

$$\varphi(m_1 m_2) = p_1^{\alpha_1 - 1}(p_1 - 1) \cdots p_k^{\alpha_k - 1}(p_k - 1)q_1^{\beta_1 - 1}(q_1 - 1) \cdots q_l^{\beta_l - 1}(q_l - 1)$$

$$= \varphi(m_1)\varphi(m_2) .$$

THEOREM 4.3. We shall first prove by induction on α that for any prime p, $a^{(p^{\alpha-1}(p-1))} - 1$ is divisible by p^{α}. For $\alpha = 1$ the assertion is simply a corollary of Fermat's little theorem, the validity of which has already been established. Thus, the basis for induction has been proved.

Let us suppose now that $p^{\alpha} | a^{p^{\alpha-1}(p-1)} - 1$, and let us consider the expression $a^{p^{\alpha}(p-1)} - 1$. We must prove that it is divisible by $p^{\alpha+1}$. However,

$$a^{p^{\alpha}(p-1)} - 1 = (a^{p^{\alpha-1}(p-1)})^p - 1.$$

Since $a^{p^{\alpha-1}(p-1)} - 1$ is by hypothesis divisible by p^{α}, the number $a^{p^{\alpha-1}(p-1)}$ has the form $Np^{\alpha} + 1$. This means that

$$a^{p^{\alpha}(p-1)} - 1 = (Np^{\alpha} + 1)^p - 1;$$

that is, by the binomial formula,

$$a^{p^{\alpha}(p-1)} - 1 = N^p p^{\alpha p} + \binom{p}{1} N^{p-1} p^{\alpha(p-1)} + \cdots + \binom{p}{p-1} N p^{\alpha} + 1 - 1.$$

In the last sum, the first term is divisible by $p^{\alpha+1}$, since it is divisible by $p^{\alpha p}$, and $\alpha p \geq \alpha + 1$. In each of the following $p - 1$ terms, the exponent of p is at least α; moreover, by the corollary to theorem 1.13, the binomial coefficient is divisible by p. This means that each of these terms is divisible by $p^{\alpha+1}$. Finally, the difference $1 - 1 = 0$ may be discarded. By theorem 1.6, therefore, $p^{\alpha+1} | a^{p^{\alpha}(p-1)} - 1$, completing the inductive step. What we now have is Euler's theorem in the case that $m = p^{\alpha}$ for some α.

Let us now suppose that Euler's theorem has been proved for the numbers m_1 and m_2, with m_1 and m_2 relatively prime. We shall prove Euler's theorem for the number $m = m_1 m_2$. Once we have done this, we may set $m_1 = p_1^{\alpha_1} \cdots p_k^{\alpha_k}$ and $m_2 = p_{k+1}^{\alpha_{k+1}}$ to obtain the inductive transition needed to establish the theorem in general. Let us therefore proceed with this proof.

Since a is relatively prime to $m = m_1 m_2$ by the hypothesis of the theorem, a must also be relatively prime to m_1, as is $a^{\varphi(m_2)}$. Therefore, by hypothesis,

$$(a^{\varphi(m_2)})^{\varphi(m_1)} - 1 = a^{\varphi(m_1)\varphi(m_2)} - 1 = a^{\varphi(m_1 m_2)} - 1 = a^{\varphi(m)} - 1$$

is divisible by m_1. In exactly the same way we see that $a^{\varphi(m)} - 1$ is divisible by m_2. And since the numbers m_1 and m_2 are relatively prime,

$a^{\varphi(m)} - 1$ is divisible by their product, that is, by m. Euler's theorem is thus proved.

THEOREM 4.4. Let

$$k_1 = \varphi(m)q_1 + r\,,$$

$$k_2 = \varphi(m)q_2 + r \quad (0 \le r < \varphi(m))\,.$$

Then

$$a^{k_1} = a^{\varphi(m)q_1 + r} = (a^{\varphi(m)})^{q_1}a^r\,.$$

On the basis of Euler's theorem and theorem 2.2, $a^{\varphi(m)q_1}a^r$ is congruent modulo m to a^r. Similarly, a^{k_2} is congruent modulo m to a^r, which means that the numbers a^{k_1} and a^{k_2} are congruent modulo m.

THEOREM 4.5. Let us first find one solution (x', y') of this equation. To do this, it is clearly sufficient to find a number x' such that $b|ax' - c$. By Euler's theorem, $b|a^{\varphi(b)} - 1$. This means that $b|ca^{\varphi(b)} - c$, and we may take the number $ca^{\varphi(b)-1}$ for x'.

Now suppose that (x'', y'') is some other solution of the equation $ax + by = c$. We shall show that the numbers x' and x'' are congruent modulo b. Indeed, let

$$ax' + by' = c\,;$$

$$ax'' + by'' = c\,.$$

Subtracting the second equation from the first, term by term, we obtain

$$a(x' - x'') + b(y' - y'') = 0\,,$$

from which $b|a(x' - x'')$. Since a and b are relatively prime by hypothesis, $b|x' - x''$ by theorem 1.12; thus, by theorem 2.1, x' and x'' are congruent modulo b.

Thus, all the desired values of x are to be found among the numbers

$$x_t = ca^{\varphi(b)-1} + bt\,.$$

But if $x_t = ca^{\varphi(b)-1} + bt$, it is a simple matter to check that $y_t = c[(1 - a^{\varphi(b)})/b] - at$ is the unique number satisfying $ax_t + by_t = c$; that is, the pairs (x_t, y_t) are exactly the solutions of our equation.

THEOREM 4.6. Since 10 and m are relatively prime, the numbers $10a + b$ and $(10a + b)10^{\varphi(m)-1}$ are—by theorem 1.15—equidivisible by m. But

$$(10a + b)10^{\varphi(m)-1} = 10^{\varphi(m)}a + 10^{\varphi(m)-1}b\,,$$

so that $10a + b$ and $a + kb$ are equidivisible by m, since by Euler's theorem, $10^{\varphi(m)}$ and 1 are congruent modulo m and by hypothesis, $10^{\varphi(m)-1}$ and k are congruent modulo m.

6

Solutions to Problems

Problem 1.1. $0 = a \cdot 0$ for any a.

Problem 1.2. $a = 1 \cdot a$ for any a.

Problem 1.3. Suppose that $a|1$. This means that $1 = ac$ for some integer c. From this it follows that $|a| \leq 1$. And since $a \neq 0$, we must have $a = \pm 1$.

Problem 1.4. It is sufficient to take any $c > 1$ and set $b = ac$.

Problem 1.5. For b we may take, for example, $2a$. Suppose that for some c, $c|2a$ and $a|c$. This means that there exist d_1 and d_2 such that $2a = d_1 c$ and $c = d_2 a$. From this it follows that $2a = d_1 d_2 a$, or, cancelling a, that

$$2 = d_1 d_2 .$$

But such an equality is possible only if one of the numbers d_1 and d_2 is equal to 1, and the other is equal to 2. If $d_1 = 1$, then $c = 2a = b$; if $d_2 = 1$, then $c = a$. The same proof clearly holds if 2 is replaced by any prime p.

Problem 1.6. The proofs differ in no substantial way from the proofs for ordinary divisibility.

Problem 1.7. Let n be some fixed number greater than 1. We shall say that $(b|a)_n$ if there exists an integer c such that $a = bc$ and $c \leq n$. The correctness of the theorems analogous to theorems 1.1, 1.3, and 1.4 may be verified without difficulty. If we take $b = na$ and $c = nb$, however, then $(a|b)_n$ and $(b|c)_n$, but $c = n^2 a$, and, since $n^2 > n$, the relation $(a|c)_n$ does not hold. In exactly the same way, the divisibility $(b|a + a)_n$ does not hold.

Problem 1.8. a. Suppose that there are two minimal elements, a_1 and a_2. By the law of dichotomy, either $a_1 \geq a_2$ or $a_2 \geq a_1$. If $a_1 \geq a_2$, then, by

50

the minimality of a_1, it follows that $a_1 = a_2$. If, however, $a_2 \geq a_1$, then $a_1 = a_2$ follows from the minimality of a_2.

b. Let a be some number, and let b_1 and b_2 be two numbers immediately preceding it. By the law of dichotomy, either $b_1 \geq b_2$ or $b_2 \geq b_1$. We may suppose without loss of generality that $b_1 \geq b_2$. We then have $a \geq b_1 \geq b_2$, and since the number b_2 immediately precedes the number a, either $b_1 = a$ or $b_1 = b_2$. But, by hypothesis, $b_1 \neq a$, so that $b_1 = b_2$, proving the desired uniqueness.

c. An immediate successor of a number a is a number b such that $b \geq a$, $b \neq a$, and from $b \geq c \geq a$ it follows that either $c = b$ or $c = a$.

Let us suppose that some number a has no immediate successor. This means that, for any $a_n \geq a$, $a_n \neq a$, there is an a_{n+1} distinct from a_n and from a so that $a_n \geq a_{n+1} \geq a$. Let us now take an arbitrary $a_1 \geq a$ and distinct from a (by property 1.2 such a number exists), and, beginning with it, construct an infinite sequence of distinct numbers

$$a_1 \geq a_2 \geq \cdots \geq a_n \geq a_{n+1} \geq \cdots \geq a\,.$$

The existence of this sequence contradicts property 1.4. Consequently, an immediate successor exists. Its uniqueness is established by using the law of dichotomy exactly as in part (b).

Problem 1.9. Transitivity (property 1.3), unboundedness of the set of numbers (property 1.5), well-ordering (property 1.4), and the existence of an immediate predecessor (property 1.6) remain valid. The law of dichotomy (property 1.7) may be replaced by the law of *trichotomy* (either $a > b$, or $a < b$, or $a = b$).

The property of reflexivity (1.1) does not hold, for $a > a$ is always false.

Finally, property 1.2 remains formally valid, for, strictly speaking, this assertion reads as follows: "For any two natural numbers a and b, from $a > b$ and $b > a$ it follows that $a = b$." Since the premise ($a > b$ and $b > a$) never holds, the implication is formally true.

Problem 1.10. Suppose that a set is ordered by a relation $\vdash$ having properties 1.1 through 1.7. As has already been shown, it has a minimal element. Let us denote this minimal element by a_0. It follows from the results of problem 1.8 that each element has an immediate successor. Let us denote the immediate successor of a_0 by a_1, the immediate successor of a_1 by a_2, and so on. As a result, we obtain a sequence

$$a_0, a_1, a_2, \ldots, \tag{6.1}$$

in which for any n, $a_{n+1} \vdash a_n$. By reflexivity and transitivity of the relation $\vdash$, it follows that $a_i \vdash a_j$ if and only if $i \geq j$. We now have only to show the sequence (6.1) contains all the objects of our set. This is accomplished by means of a fairly subtle argument by induction.

Let b_0 be an element of our ordered set. We shall construct inductively a decreasing sequence of objects all smaller than b_0 as follows: Suppose b_{n-1} has already been chosen, and compare it with a_0. If $b_{n-1} = a_0$, the sequence is complete. If $b_{n-1} \neq a_0$, then take b_n to be the immediate predecessor of b_{n-1}. This process results in a sequence of distinct elements:

$$b_0 \vdash b_1 \vdash b_2 \vdash \cdots \vdash b_n \vdash \cdots\,.$$

On the basis of property 1.4, this sequence must have a last element. But from the very rule of construction of this sequence, its last element can only be a_0. We may suppose without loss of generality that $b_n = a_0$.

It is not difficult to verify that if some number a immediately precedes b, then b immediately follows a. This means that $b_{n-1} = a_1$, $b_{n-2} = a_2$, ..., $b_0 = a_n$.

This last equality shows that any element b_0 of our set belongs to sequence (6.1).

Problem 1.11. Let a be some number. We shall call any sequence of distinct numbers $a_0 = a, a_1, a_2, \ldots, a_n$, for which

$$a_0 \mathrel{\text{\reflectbox{$\sqsubset$}}} a_1 \mathrel{\text{\reflectbox{$\sqsubset$}}} a_2 \mathrel{\text{\reflectbox{$\sqsubset$}}} \cdots \mathrel{\text{\reflectbox{$\sqsubset$}}} a_n , \qquad (6.2)$$

where a_n is minimal under the ordering $\mathrel{\text{\reflectbox{$\sqsubset$}}}$, a *chain of predecessors* of a_0. The number n will be called the *length* of this chain.

We shall show first that, under the conditions we have imposed on the order $\mathrel{\text{\reflectbox{$\sqsubset$}}}$, no number can have arbitrarily long chains of predecessors.

Indeed, let a be some number, and let $b_1, b_2, \ldots, b_k$; be numbers that immediately precede it.

If a_1 does not immediately precede a_0, we may, on the basis of property 1.9, insert in (6.2) a number immediately preceding a (recall $a = a_0$). Consequently, if there are arbitrarily long chains of predecessors of a, there must also be arbitrarily long chains of its predecessors that begin with numbers immediately preceding a. In the future, we shall consider only such chains.

Each of these chains of predecessors of a is longer by exactly one than some chain of predecessors of one of the numbers immediately preceding a. If each of them had chains of predecessors of bounded length, then a itself could not have arbitrarily long chains of predecessors.

This means that, under our assumption, at least one of the numbers immediately preceding a_0 has arbitrarily long chains of prececessors. Let us denote this number by a_1 and apply to it all the arguments just given with regard to a_0. This gives us some number a_2 that immediately precedes a_1 and has arbitrarily long chains of predecessors. Repeating this process, we arrive at a sequence

$$a_0 \mathrel{\text{\reflectbox{$\sqsubset$}}} a_1 \mathrel{\text{\reflectbox{$\sqsubset$}}} a_2 \mathrel{\text{\reflectbox{$\sqsubset$}}} \cdots ,$$

which, by virtue of property 1.4, must break off sooner or later. This means that the sequence will have an element to which our arguments will no longer apply. But we have already established the applicability of these arguments to each element of the sequence. This contradiction shows that no number has arbitrarily long chains of predecessors.

For each number a, therefore, it is possible to choose a longest from among its chains of predecessors. Let us denote its length by $n(a)$. If b immediately precedes a, then clearly $n(b) \le n(a) - 1$, and, for all minimal a, $n(a) = 0$.

Suppose, finally, that $A(a)$ is an assertion depending on a. Let us denote by $B(n)$ the assertion "$A(a)$ is valid for all objects a for which $n(a) = n$."

Then, as may easily be seen, the formulation of the principle of induction in the new form for the assertions $A(a)$ coincides with the formulation of this principle in the old form for the assertions $B(n)$.

Problem 1.12. For any even integers a and b, there exist even numbers q and r such that

$$a = bq + r \qquad (0 \le r < 2b). \tag{6.3}$$

These numbers q and r are unique.

Proof. By theorem 1.7, there exist q_0 and r_0 such that

$$a = bq_0 + r_0 \qquad (0 \le r_0 < b).$$

Because of theorem 1.6, the number r_0 must be even. If q_0 is also even, we may set $q = q_0$ and $r = r_0$. But if the number q_0 is odd, we take $q = q_0 - 1$ (which is even), and $r = r_0 + b$. In either case, we obtain relation (6.3).

Let us suppose that, in addition to (6.3), there is another representation

$$a = bq' + r' \qquad (0 \le r' < 2b),$$

in which the numbers q' and r' are even. Then we have

$$b(q' - q) = r - r'.$$

Since

$$0 \le |r - r'| < 2b,$$

we must have $|q' - q| < 2$. But $q' - q$ is an even number, which means that $q' - q = 0$, from which uniqueness follows at once.

Problem 1.13. Let p be the smallest prime divisor of the number a. From this it follows that $a = pb$. Any prime divisor q of the number b is, at the same time, a divisor of a. Therefore, $q \ge p$, and so $b \ge p$, so that $a \ge p^2$, and, finally, $p \le \sqrt{a}$. No better bound than this is possible, as may be seen by considering the number $a = p^2$.

Problem 1.14. Necessity. Suppose that $b|a$. From theorem 1.13, it follows that each prime divisor of b is a prime divisor of a. Thus, b has the form

$$p_1^{\beta_1} p_2^{\beta_2} \cdots p_r^{\beta_r}, \tag{6.4}$$

where $0 \le \beta_1, 0 \le \beta_2, \ldots, 0 \le \beta_r$. Let us suppose that some $\beta_i > \alpha_i$. Without loss of generality, we may assume $\beta_1 > \alpha_1$. Set $a = bc$. Then

$$p_1^{\alpha_1} p_2^{\alpha_2} \cdots p_r^{\alpha_r} = p_1^{\beta_1} p_2^{\beta_2} \cdots p_r^{\beta_r} c.$$

Cancelling $p_1^{\alpha_1}$ on both sides of this equation, we have

$$p_2^{\alpha_2}\cdots p_r^{\alpha_r} = p_1^{\beta_1-\alpha_1}p_2^{\beta_2}\cdots p_r^{\beta_r}c\ .$$

Since $\beta_1 > \alpha_1$ implies $\beta_1 - \alpha_1 > 0$, p_1 must divide the right side of the new equation; but p_1 does not divide the left side. Since this is a contradiction, we are forced to conclude that $\beta_i \leq \alpha_i$ for each $i = 1$, $2, \ldots, r$, completing the proof of necessity.

Sufficiency. To prove the sufficiency we need only note that if b has the indicated form, then

$$a = bp_1^{\alpha_1-\beta_1}p_2^{\alpha_2-\beta_2}\cdots p_r^{\alpha_r-\beta_r}\ .$$

Problem 1.15. As we have already shown, each divisor of the number a with canonical decomposition $p_1^{\alpha_1}p_2^{\alpha_2}\cdots p_r^{\alpha_r}$ must have form (6.4), where β_1 assumes $\alpha_1 + 1$ values: $0, 1, 2, \ldots, \alpha_1$; β_2 assumes $\alpha_2 + 1$ values, and so on. Since any combination of these values is possible and the set of all such combinations gives us all the divisors of a, with each occurring exactly once (if any divisor were repeated more than one time, then it would have more than one canonical decomposition), the number of divisors of a is

$$(\alpha_1 + 1)(\alpha_2 + 1)\cdots(\alpha_r + 1)\ .$$

Problem 1.16. Suppose that the canonical decomposition of a is $p_1^{\alpha_1}p_2^{\alpha_2}$ $\cdots p_k^{\alpha_k}$. Clearly, we may set $p_1 = 2$, $\alpha_1 \geq 2$ and $p_2 = 3$, $\alpha_2 \geq 1$. Furthermore, we have

$$(\alpha_1 + 1)(\alpha_2 + 1)\cdots(\alpha_k + 1) = 14\ .$$

Hence, $k = 2$, $\alpha_1 + 1 = 7$, and $\alpha_2 + 1 = 2$ (since $\alpha_1 + 1 = 2$ implies that $4 \nmid a$). Thus, $a = 2^6 \cdot 3 = 192$.

Problem 1.17. We have

$$\tau(a^2) = \tau(p_1^{2\alpha_1}p_2^{2\alpha_2}) = (2\alpha_1 + 1)(2\alpha_2 + 1) = 81\ ,$$

so that $(2\alpha_1 + 1)(2\alpha_2 + 1)$ is a decomposition of 81 into two factors. Since the numbering of the prime divisors of a is arbitrary, we need only consider the following possibilities:

$$\begin{aligned}
2\alpha_1 + 1 &= 1\ , & 2\alpha_2 + 1 &= 81\ ; \\
2\alpha_1 + 1 &= 3\ , & 2\alpha_2 + 1 &= 27\ ; \\
2\alpha_1 + 1 &= 9\ , & 2\alpha_2 + 1 &= 9\ .
\end{aligned}$$

In the first of these cases, $\alpha_1 = 0$, which contradicts the assumption that $\alpha > 0$. The remaining cases give us

$$\begin{aligned}
\alpha_1 &= 1\ , & \alpha_2 &= 13\ ; \\
\alpha_1 &= 4\ , & \alpha_2 &= 4\ .
\end{aligned}$$

This means that either

$$\tau(a^3) = \tau(p_1^{3\alpha_1} p_2^{3\alpha_2}) = \tau(p_1^3 p_2^{39}) = (3 + 1)(39 + 1) = 160,$$

or

$$\tau(a^3) = \tau(p_1^{3\alpha_1} p_2^{3\alpha_2}) = \tau(p_1^{12} p_2^{12}) = 13 \cdot 13 = 169.$$

Problem 1.18. Let $p_1^{\alpha_1} p_2^{\alpha_2} \cdots p_k^{\alpha_k}$ be the canonical decomposition of the number a. The condition of the problem gives us

$$p_1^{\alpha_1} p_2^{\alpha_2} \cdots p_2^{\alpha_k} = 2(\alpha_1 + 1)(\alpha_2 + 1) \cdots (\alpha_k + 1),$$

or

$$\frac{p_1^{\alpha_1}}{\alpha_1 + 1} \frac{p_2^{\alpha_2}}{\alpha_2 + 1} \cdots \frac{p_k^{\alpha_k}}{\alpha_k + 1} = 2. \tag{6.5}$$

We note that

$$\frac{2^1}{1 + 1} = 1 < \frac{2^2}{2 + 1} < \frac{2^3}{3 + 1} = 2 < \frac{2^\alpha}{\alpha + 1} \qquad (\alpha \geq 4),$$

$$1 < \frac{3^1}{1 + 1} < 2 < \frac{3^\alpha}{\alpha + 1} \qquad (\alpha \geq 2),$$

$$2 < \frac{p^\alpha}{\alpha + 1} \qquad (p \geq 5, \alpha \geq 1).$$

Therefore, each fraction on the left side of (6.5) is not smaller than 1, and, consequently, none of the fractions may be larger than 2. This means that only fractions from the following collection can appear on the left side of (6.5):

$$\frac{2^1}{1 + 1}, \frac{2^2}{2 + 1}, \frac{2^3}{3 + 1}, \frac{3^1}{1 + 1};$$

and the product of the fractions must be 2. But this can be so in only two cases: when the only fraction on the left side of (6.5) is $2^3/(3 + 1)$, or when only the two fractions $2^2/(2 + 1)$ and $3^1/(1 + 1)$ appear there. The two answers to the problem—8 and 12—correspond to these two cases.

Problem 1.19. The analogues of theorems 1.11–1.14 are false for even divisibility.

Indeed, the numbers 30 and 42 are even primes (that is, they are not products of any two even numbers). Their least common even multiple is 420 and their product is 1260.

Furthermore, $60 = 6 \cdot 10$ is evenly divisible by the even prime 30; 6 and 30 are evenly relatively prime, but 10 is not evenly divisible by 30.

Finally, the decompositions $60 = 6 \cdot 10$ and $60 = 30 \cdot 2$ are two distinct factorizations of the number 60 into even primes.

Problem 1.20. Let $p_1, p_2, \ldots, p_k$ be a complete list of all the primes entering into the canonical decomposition of at least one of a and b. Let us set

$$a = p_1{}^{\alpha_1} p_2{}^{\alpha_2} \cdots p_k{}^{\alpha_k},$$

$$b = p_1{}^{\beta_1} p_2{}^{\beta_2} \cdots p_k{}^{\beta_k}.$$

(If a is not divisible by p_i, then $\alpha_i = 0$; if b is not divisible by p_i, then $\beta_i = 0$.) Let γ_i be the larger of the two numbers α_i and β_i for each $i = 1, 2, \ldots, k$, and let δ_i be the smaller of them.

Then, because of our result in problem 1.14, the greatest common divisor of a and b is $p_1{}^{\delta_1} p_2{}^{\delta_2} \cdots p_k{}^{\delta_k}$, and their least common multiple is $p_1{}^{\gamma_1} p_2{}^{\gamma_2} \cdots p_k{}^{\gamma_k}$.

Problem 2.1. a. 116 is congruent to 4 and 17 is congruent to 1 modulo 8. This means that A is congruent to $5^{21} = (5^2)^{10} \cdot 5$. But $5^2 = 25$ is congruent to 1 modulo 8. Consequently, A gives a remainder of 5 on division by 8.

b. It is clear that $14^{256} \equiv 196^{128} \equiv 9^{128} \equiv 81^{64} \equiv (-4)^{64} \equiv 4^{64} \equiv 16^{32} \equiv (-1)^{32} \equiv 1 (\mathrm{mod}\ 17)$.

Problem 2.2. a. Let n_1 be the remainder on division of n by 6. Then n_1 may assume the values 0, 1, 2, 3, 4, and 5, and $n_1{}^3 + 11n_1$ is congruent modulo 6 to $n^3 + 11n$. This means that we need test only the numbers 0, 12, 30, 60, 108, and 180 for divisibility by 6. But all of these numbers are divisible by 6, and so $n^3 + 11n$ will be divisible by 6 for any n.

b. For $n \geq 2$, we have (using the binomial theorem):

$$4^n + 15n - 1 = (3 + 1)^n + 15n - 1$$

$$= 3^n + 3^{n-1} \binom{n}{1} + \cdots + 3^2 \binom{n}{n-2}$$

$$+ 3 \binom{n}{n-1} + 1 + 15n - 1$$

$$= 9 \left(3^{n-2} + 3^{n-3} \binom{n}{1} + \cdots + \binom{n}{n-2} \right)$$

$$+ 18n,$$

which is clearly divisible by 9

For $n = 1$, the expression is equal to $4^1 + 15 \cdot 1 - 1 = 18$, which is also divisible by 9, completing the proof.

c. The proof is by induction.
For $n = 0$,

$$10^{(3^0)} - 1 = 10^1 - 1 = 9 \quad \text{and} \quad 3^{0+2} = 9,$$

so $3^{n+2} | 10^{(3^n)} - 1$.

Suppose now that the divisibility

$$3^{n+2} | 10^{(3^n)} - 1$$

is known to be valid. Then

$$10^{(3^{n+1})} - 1 = (10^{(3^n)})^3 - 1^3$$
$$= (10^{(3^n)} - 1)(10^{2 \cdot (3^n)} + 10^{(3^n)} + 1).$$

The first factor on the right is divisible by 3^{n+2} by the induction hypothesis. In the second factor, we may replace the tens by ones, since 10 is congruent to 1 modulo 3. The resulting sum, 3, is divisible by 3, and, therefore, so is the second factor itself. Consequently, the whole product is divisible by $3^{n+3} = 3^{(n+1)+2}$, which is what is required.

d. Clearly, a^2 is congruent to $a - 1$ modulo $a^2 - a + 1$. This means that $a^{2n+1} + (a - 1)^{n+2}$ is congruent to

$$a^{2n+1} + (a^2)^{n+2} = a^{2n+1} + a^{2n+4}$$
$$= a^{2n+1}(1 + a^3)$$
$$= a^{2n+1}(1 + a)(1 - a + a^2),$$

so $a^2 - a + 1 | a^{2n+1} + (a - 1)^{n+2}$, as was to be shown.

Problem 2.3. Let $\sim$ be an equivalence relation on a set of numbers. Let us take an arbitrary number a and consider all numbers equivalent to a. At least one such number, namely a, is known to exist because the relation $\sim$ is reflexive. Any two of these numbers are equivalent to one another by virtue of the transitivity of the relation $\sim$. We shall denote the class of these numbers by K.

Now let us consider an arbitrary number b not belonging to K. If $b \sim c$ were to hold for some number c in K, then $b \sim a$ would also hold, but this cannot be, by the choice of b. This means that none of the numbers lying outside K is equivalent to any of the numbers in K. Consequently, K is an equivalence class containing a.

Finally, since the number a was completely arbitrary, the arguments given show that each number belongs to some equivalence class, and so the proof is complete.

Problem 2.4. Clearly, among the $m + 1$ numbers $0, 1, \ldots, m$, there are two that belong to the same class, since $\sim$ splits the numbers into only m classes. Suppose that these numbers are k and l: $k \sim l$. Since, generally speaking, there may be more than one such pair of numbers, let us choose that pair for which the quantity $|k - l|$ is greatest. Since $-l \sim -l$, we obtain:

$$k - l \sim l - l = 0 .$$

Furthermore, we find that, for any integer n,

$$n(k - l) \sim 0 .$$

Finally, for any r,

$$n(k - l) + r \sim r ;$$

that is, from $a \equiv b(\text{mod } k - l)$ it follows that $a \sim b$. Thus, each equivalence class of the relation $\sim$ contains some residue class modulo $k - l$.

In order to have m equivalence classes under $\sim$, it is therefore necessary to have $k - l \geq m$. But by our construction, $k - l \leq m$. Thus, $k - l = m$ and each equivalence class under $\sim$ contains exactly 1 residue class modulo m.

Problem 2.5. Two such rules are the following:

a. Both sides of a congruence and the modulus may be divided by the same number (distinct from zero, of course).

Indeed,

$$ad \equiv bd(\text{mod } md)$$

means that

$$md \,|\, ad - bd = (a - b)d ;$$

that is, that $m|a - b$, and hence, $a \equiv b(\text{mod } m)$.

b. Both sides of a congruence may be divided by a number that is relatively prime to the modulus.

Indeed, if d and m are relatively prime, then from

$$ad \equiv bd(\text{mod } m) ,$$

that is, from $m|(a - b)d$, it follows on the basis of theorem 1.12 that $m|a - b$, as was required.

Problem 2.6. Let us suppose, on the contrary, that

$$ka \equiv la(\text{mod } p) , \qquad 1 \leq k < l \leq p - 1 .$$

This means that $p|(l - k)a$. Since p does not divide a, we must have $p|l - k$. But this is impossible, since $0 < l - k < p$.

Problem 2.7. Necessity. Suppose that the number p is prime. Let us choose any q such that $0 < q < p$. Among the numbers $q, 2q, \ldots, (p - 1)q$, there is exactly one that leaves a remainder of 1 on division by p. Suppose that this number is $\bar{q}q$:

$$\bar{q}q \equiv 1(\bmod p)\,. \tag{6.6}$$

On the other hand, among the numbers $\bar{q}, 2\bar{q}, \ldots, (p - 1)\bar{q}$ there is also exactly one that leaves a remainder of 1 on division by p. This, as has already been established, is the number $q\bar{q}$.

Let us clarify the circumstances under which $q = \bar{q}$. In all such cases, the congruence (6.6) may be rewritten as

$$q^2 \equiv 1(\bmod p)\,,$$

or, equivalently, as

$$q^2 - 1 \equiv 0(\bmod p)\,.$$

This means that

$$p|q^2 - 1 = (q + 1)(q - 1)\,.$$

Since p is prime, theorem 1.13 shows that either $p|q + 1$ or $p|q - 1$. Since the number q lies between 0 and p, the first case is possible only for $q = p - 1$ and the second for $q = 1$.

Consequently, given any q different from 1 and $p - 1$, $q \neq \bar{q}$; so the remaining numbers $2, \ldots, p - 2$ may be paired in such a way that the product of the numbers forming each pair leaves a remainder of 1 on division by p. Let us write out the congruences of the form (6.6) for all such pairs and add to this list the congruence

$$p - 1 \equiv p - 1(\bmod p)\,;$$

and then let us multiply all these $(p - 1)/2$ congruences term by term.

As the result of this multiplication, we obtain on the left side the product of all the numbers from 2 to $p - 1$ and, on the right side, $p - 1$:

$$2 \cdot 3 \cdots (p - 1) \equiv p - 1(\bmod p)\,,$$

or

$$1 \cdot 2 \cdot 3 \cdots (p - 1) + 1 \equiv 0(\bmod p)\,,$$

which means simply that

$$p|1 \cdot 2 \cdots (p - 1) + 1\,,$$

as was required.

Sufficiency. If the number p is not prime, then it may be decomposed into two smaller factors: $p = p_1 p_2$.

If $p_1 \neq p_2$, then both p_1 and p_2 occur as factors in the product $1 \cdot 2 \cdots (p - 1)$, which is therefore divisible by $p_1 p_2$, that is, by p. If, however, $p_1 = p_2 = q$, then $p = q^2$ (that is, p is a perfect square). If $q > 2$, then $p > 2q$ and both the numbers q and $2q$ enter into the product $1 \cdot 2 \cdots (p - 1)$, so that this product is divisible by $q^2 = p$. In either case, $1 \cdot 2 \cdots (p - 1) + 1$ cannot be divisible by p. Finally, if $q = 2$, so that $p = 4$, then $1 \cdot 2 \cdot 3 + 1 = 7$, which is not divisible by 4.

Problem 2.8. Theorem. Let $m = p_1^{\alpha_1} p_2^{\alpha_2} \cdots p_k^{\alpha_k}$ be the canonical decomposition of m. Then for the numbers A and B to be congruent modulo m, it is necessary and sufficient that they be congruent modulo $p_i^{\alpha_i}$ $(i = 1, 2, \ldots, k)$.

Proof. For A and B to be congruent modulo m, it is necessary and sufficient that $m | A - B$. But by theorem 1.16, this is equivalent to

$$p_1^{\alpha_1} | A - B, \ldots, p_k^{\alpha_k} | A - B,$$

that is, to

$$A \equiv B (\operatorname{mod} p_i^{\alpha_i}), \qquad i = 1, 2, \ldots, k.$$

Problem 3.1. a. Application of the Euclidean algorithm to the numbers a and b yields a series of equations of the form:

$$
\begin{aligned}
a &= b q_0 + r_1, \\
b &= r_1 q_1 + r_2, \\
r_1 &= r_2 q_2 + r_3, \\
&\cdots\cdots\cdots\cdots\cdots \\
r_{n-2} &= r_{n-1} q_{n-1} + r_n, \\
r_{n-1} &= r_n q_n.
\end{aligned}
\tag{6.7}
$$

We have $r_n | r_{n-1}$. This, together with $r_{n-2} = r_{n-1} q_{n-1} + r_n$, gives us $r_n | r_{n-2}$. Arguing similarly, we may move up the system of equations (6.7), until we finally conclude that $r_n | a$ and $r_n | b$. This means that r_n is a common divisor of a and b.

Let d be any common divisor of a and b. This fact, together with the equation $a = b q_0 + r_1$, gives us $d | r_1$. Moving in this way down the system of equalities (6.7), we find successively that $d | r_2$, $d | r_3$, ..., $d | r_n$. This means that r_n is divisible by any common divisor of a and b and is thus the greatest common divisor of these numbers.

b. The proof is by induction and uses the Euclidean algorithm applied to a and b. Setting $A_0 = 0$, $B_0 = 1$, $A_1 = 1$, $B_1 = -q_0$, we have $r_0 = b = a A_0 + b B_0$ and $r_1 = a A_1 + b B_1$. Suppose now that

$$r_{k-1} = A_{k-1} a + B_{k-1} b,$$

$$r_k = A_k a + B_k b.$$

Then

$$r_{k+1} = r_{k-1} - r_k q_k = (A_{k-1} - q_k A_k)a + (B_{k-1} - q_k B_k)b \,,$$

and we have only to set

$$A_{k+1} = A_{k-1} - q_k A_k \,,$$

$$B_{k+1} = B_{k-1} - q_k B_k \,,$$

in order to have $r_{k+1} = A_{k+1}a + B_{k+1}b$. Then clearly, by this construction, the numbers A_n and B_n will be the desired A and B.

Problem 3.2. We will consider only theorem 1.12 as an illustration. The proofs of the other theorems use similar techniques.

Suppose that $c|ab$. If b and c are relatively prime, then, by the preceding problem, we may find integers B and C such that

$$bB + cC = 1 \,,$$

or, after multiplication by a,

$$abB + acC = a \,;$$

but $c|ab$ by hypothesis and since $c|ac$, we have $c|a$, as desired.

Problem 3.3. We shall limit ourselves to considering a criterion for congruence modulo 8.

Suppose that an arbitrary natural number A is represented in the form $1000a + b$, where $0 \le b < 1000$ (that is, b is simply the last three digits of A). Let

$$f(A) = \begin{cases} b \text{ if } A \ge 1000 \,, \\ \text{the remainder on division of } A \text{ by } 8 \text{ if } 8 \le A < 1000 \,, \\ \text{undefined if } A < 8 \,. \end{cases}$$

The one may easily verify that the function f is the required criterion for congruence.

Problem 3.4. For those having a canonical decomposition of the form $2^\alpha 5^\beta$ for some α and β.

Problem 3.5. Conditions (a) and (b) are automatically satisfied. Since the numbers 10 and 1 (and, therefore, 10^2 and 1, 10^3 and 1, and so on) are congruent modulo 3, so are the numbers A and $f(A)$. Finally, the fact that $f(A) < A$ for $A \ge 3$, is established by a simple calculation.

Problem 3.6. a. $f_2(858,773) = 38$; $f_2(38) = 11$; $f_2(11) = 2$. $f_2(789,988)$ $= 49$; $f_2(49) = 13$; $f_2(13) = 4$; $f_2(4) = 1$.

b. $f_2(A) = 4444 \cdot 4 = 17,776$; $f_2(17,776) = 28$; $f_2(28) = 10$; $f_2(10) = 1$.

Problem 3.7. The criterion for congruence modulo 9 is exactly analogous to the criterion for congruence modulo 3.

To obtain a criterion for congruence on division by 11, we represent the number A in the form

$$A = 10^{2n}a_n + 10^{2n-2}a_{n-1} + \cdots + 10^2 a_1 + a_0,$$

where $0 \le a_i < 100$. Such a representation clearly corresponds to simply breaking up the number into two-place "blocks" (from right to left). Let

$$f(A) = \begin{cases} a_0 + a_1 + \cdots + a_n \text{ if } A \ge 100, \\ \text{the remainder on division of } A \text{ by 11 if } 11 \le A < 100, \\ \text{not defined if } A < 11. \end{cases}$$

It is not hard to show that the numbers A and $f(A)$ are indeed congruent modulo 11 and that, in addition, $f(A) < A$.

Another criterion for congruence on division by 11 is obtained by representing the number A in the form

$$A = 10^n a_n + 10^{n-1}a_{n-1} + \cdots + 10a_1 + a_0$$

and using the fact that 10 is congruent to -1 and 100 is congruent to 1 modulo 11. Then A is congruent to the number $a_0 - a_1 + a_2 - a_3 + \cdots \pm a_n$, and one may easily formulate a criterion for congruence on this basis.

Finally, we may break up the number A into 3-place "blocks" and represent it in the form

$$10^{3n}a_n + 10^{3n-3}a_{n-1} + \cdots + 10^3 a_1 + a_0$$

$(0 \le a_i < 1000)$. Then, modulo 37, A is congruent to the sum $a_0 + a_1 + \cdots + a_n$, and, modulo 7, 11, and 13, to the alternating sum $a_0 - a_1 + a_2 - \cdots \pm a_n$.

Problem 3.8. If the numbers a and b are congruent, then $m|a - b$. Therefore, by virtue of theorem 1.6, the numbers a and b are simultaneously divisible or not divisible by m.

Four and 5 are equidivisible on division by 3, but not congruent modulo 3.

Problem 3.9. Suppose that congruence modulo m implies equidivisibility by m. Then all numbers not divisible by m leave the same remainder on division by m. Therefore, either there are no numbers which are not divisible by m (that is, $m = 1$), or there is only one non-zero remainder on division by m (that is, $m = 2$).

Problem 3.10. The relation of equidivisibility is clearly reflexive (any number is equidivisible by m with itself), symmetric (if a is equidivisible with b, then b is equidivisible with a), and transitive (if a is equidivisible with b and b is equidivisible with c, then a is equidivisible with c).

Consequently, equidivisibility is an equivalence relation. All numbers that are divisible by m fall into one class, and all numbers that are not divisible by m fall into another class.

Problem 3.11. It is not difficult to verify that for $m > 2$, equidivisibility of sums does not follow from the equidivisibility of their terms.

A necessary and sufficient condition for the equidivisibility of products to follow from the equidivisibility of their terms is that the number m be prime.

Indeed, if one of the products is divisible by a prime p, then, by theorem 1.13, at least one of the factors of this product must be divisible by p; but then p would divide the factor in the other product that is equidivisible with this factor in the first product, and hence, p would divide the second product. So p divides one product if and only if p divides both products; that is, the two products are equidivisible by p.

Conversely, if the number p is composite, then products of equidivisible factors may fail to be equidivisible. It is sufficient to set $p = p_1 p_2$ ($p_1 \neq 1, p_2 \neq 1$). Then the numbers 1 and p_1, as well as the numbers 1 and p_2, are equidivisible by p, while their products clearly are not.

The corollary does not hold, as 2 is equidivisible with 4 on division by 8, but 2^2 and 4^2 are not equidivisible by 8.

Problem 3.12. This is an immediate corollary of the first part of problem 3.8.

Problem 3.13. a. Clearly, it is enough to check that $a - 2b > 0$ whenever $A \geq 190$. Now if $a - 2b \leq 0$, then $a \leq 2b$. Since the maximum possible value for b is 9, this cannot happen if $a > 18$. But $a > 18$ whenever $A \geq 190$.

b. is clear.

c. If $A \geq 190$, we must show that $10a + b > a - 2b$, or $9a + 3b > 0$. But this is always true, since a is positive and b is nonnegative.

If $7 \leq A < 190$, (c) is clear.

Finally, $10a + b$ and $50a + 5b$ are equidivisible by 7 (since the numbers 5 and 7 are relatively prime) and so, therefore, are $10a + b$ and $50a + 5b - 7(7a + b) = a - 2b$.

Problem 3.14. 191 is congruent to 2 modulo 7, while $f_3(191) = 19 - 2 \cdot 1 = 17$ is congruent to 3 modulo 7.

Problem 3.15. Condition (c). To verify that $f(A) < A$, it is sufficient to show that $a + 4b < 10a + b$; that is $3b < 9a$. Therefore, for $a \geq 4$ (i.e., $A \geq 40$), the necessary condition is satisfied.

Condition (d*). Since 4 and 13 are relatively prime, $10a + b$ is equidivisible by 13 with $40a + 4b$, and this last number is equidivisible with $a + 4b$.

Problem 3.16. The criterion for divisibility would no longer be determinate, as $f_4(39)$ would be 39.

Problem 3.17. Suppose that we are to construct a criterion for divisibility by some m. We choose the smallest number s satisfying $m|10s + 1$ (for $m = 7$; s turned out to be 2) or $m|10s - 1$ (for $m = 13$, $s = 4$).

In the first of these cases, $A = 10a + b$ is equidivisible by m with

$$10as + bs = (10s + 1)a - a + bs$$

(since s and m must be relatively prime); that is, with $a - bs$, and in the second case, with

$$10as + bs = (10s - 1)a + a + bs ;$$

that is, with $a + bs$.

In connection with the above, the number $10a + b$

is equidivisible with $a - 5b$ on division by 17 ,
is equidivisible with $a + 2b$ on division by 19 ,
is equidivisible with $a + 7b$ on division by 23 ,
is equidivisible with $a + 3b$ on division by 29 ,
is equidivisible with $a - 3b$ on division by 31 .

The precise formulation of these criteria for divisibility is left to the reader.

Problem 3.18. a. Since 100 is congruent to 2 modulo 49, any number of the form

$$10^{2n}a_n + 10^{2n-2}a_{n-1} + \cdots + 10^2 a_1 + a_0 \qquad (0 \le a_i < 100)$$

is in fact congruent, modulo 49, to

$$2^n a_n + 2^{n-1} a_{n-1} + \cdots + 2a_1 + a_0 .$$

b. $10a + b$ is equidivisible with $a + 5b$ by 49.

Problem 3.19. Conditions (a) and (b) are automatically satisfied. Conditions (c) and (d) hold because in order to obtain $f(A)$ from A for $10^n \ge m$, one need only replace certain numbers whose sum is equal to A by their remainders on division by m (which are smaller than but congruent to the numbers themselves), thus obtaining as a new sum a smaller number which is congruent to A by theorem 1.6.

Problem 3.20. a. $r_1 = r_2 = \cdots = r_n = 0$; that is, $r_k = 0$ for $k \ge 1$;
b. $r_2 = r_3 = \cdots = r_n = 0$; that is, $r_k = 0$ for $k \ge 2$;
c. $r_1 = r_2 = \cdots = r_n = 1$; that is, $r_k = 1$ for all k;
d. $r_1 = r_3 = \cdots = r_{2t-1} = -1$; $r_2 = r_4 = \cdots = r_{2t} = 1$; that is, $r_k = (-1)^k$ for all k;
e. $r_{6t+1} = 3, r_{6t+2} = 2, r_{6t+3} = 6, r_{6t+4} = 4, r_{6t+5} = 5, r_{6t} = 1$, all t.

Problem 3.21. This is left to the reader.

Problem 4.1. Neither $2^4 - 2$ nor $2^3 - 1$ is divisible by 4.

Problem 4.2. If $p|a$, then $p|a^p - a$, and the theorem is proved. If, however, p does not divide a, then a and p are relatively prime, and the congruence in the theorem is equivalent to the congruence

$$a^{p-1} \equiv 1 (\bmod\ p) .$$

To prove that this last congruence holds, we divide each number of the form ta ($t = 1, 2, \ldots, p - 1$) by p with remainder:

$$ta = q_t p + r_t .$$

This may be rewritten as

$$\left.\begin{array}{l} a \equiv r_1 (\bmod\ p) , \\ 2a \equiv r_2 (\bmod\ p) , \\ \cdots\cdots\cdots\cdots\cdots\cdots\cdots \\ (p - 1)a \equiv r_{p-1} (\bmod\ p) . \end{array}\right\} \qquad (6.8)$$

From the result of problem 2.6, it follows that each of the numbers $1, 2, \ldots, p-1$ occurs exactly once among the numbers $r_1, \ldots, r_{p-1}$. Multiplying together all the congruences in (6.8) term by term, we obtain

$$1 \cdot 2 \cdots (p-1)a^{p-1} \equiv 1 \cdot 2 \cdots (p-1)(\bmod p).$$

The proof is completed cancelling $1 \cdot 2 \cdots (p-1)$ on both sides of the congruence.

Problem 4.3.

$$\varphi(12) = \varphi(2^2 \cdot 3) = 2^{2-1}(3-1) = 2 \cdot 2 = 4,$$
$$\varphi(120) = \varphi(2^3 \cdot 3 \cdot 5) = 2^{3-1}(3-1)(5-1) = 4 \cdot 2 \cdot 4 = 32,$$
$$\varphi(1000) = \varphi(2^3 \cdot 5^3) = 2^{3-1}5^{3-1}(5-1) = 4 \cdot 25 \cdot 4 = 400.$$

Problem 4.4. Let $m = p_1^{\alpha_1}p_2^{\alpha_2}\cdots p_k^{\alpha_k}$. Then

a. $p_1^{\alpha_1-1}(p_1-1)p_2^{\alpha_2-1}(p_2-1)\cdots p_k^{\alpha_k-1}(p_k-1) = 10.$

The product on the left must be divisible by 5. This means that either one of the numbers $p_1, p_2, \ldots, p_k$ is 5 (to be definite, let us suppose that $p_1 = 5$) or one of the differences $p_1-1, p_2-1, \ldots, p_k-1$ is divisible by 5 (in this case, suppose that $5|p_1-1$). In the first of these cases, $p_1-1 = 4$, which cannot be, since 10 is not divisible by 4. Thus we are reduced to the second case, which is possible only for $p_1 = 11$, since p_1 must be prime, and $(p_1-1)|10$. But then $\alpha_1 = 1$, and from theorem 4.2 it follows that $\varphi(m/11) = 1$; that is, either $m/11 = 1$ or $m/11 = 2$; that is, either $m = 11$ or $m = 22$. It is easy to verify that $\varphi(11) = \varphi(22) = 10$.

b. $p_1^{\alpha_1-1}(p_1-1)p_2^{\alpha_2-1}(p_2-1)\cdots p_k^{\alpha_k-1}(p_k-1) = 8.$

If m is odd, then $\alpha_1 = \alpha_2 = \cdots = \alpha_k = 1$ (since the right side of the equality above is a power of two):

$$(p_1-1)(p_2-1)\cdots(p_k-1) = 8.$$

This is possible only for $k = 2, p_1 = 3$, and $p_2 = 5$; that is, for $m = 15$.

Suppose now that the number m is even. In order to be definite, let us set $p_1 = 2$. Clearly, as before, $\alpha_2 = \cdots = \alpha_k = 1$, and we have

$$2^{\alpha-1}(p_2-1)\cdots(p_k-1) = 8.$$

Clearly, $\alpha \leq 4$. If $\alpha = 1$, then the case is exactly like the one just considered: The equality above is possible only for $k = 3$, $p_2 = 3$, and $p_3 = 5$, that is, for $m = 30$. If $\alpha = 2$, then $k = 2$, $p_2 = 5$, and $m = 20$. If $\alpha = 3$, then $k = 2$, $p_2 = 3$, and $m = 24$. Finally, if $\alpha = 4$, then $k = 1$ and $m = 16$.

Thus, the solutions to our problem are: $m_1 = 15$, $m_2 = 30$, $m_3 = 20$, $m_4 = 24$, $m_k = 16$.

Problem 4.5. Let us suppose that

$$p_1^{\alpha_1-1}(p_1 - 1)p_2^{\alpha_2-1}(p_2 - 1)\cdots p_k^{\alpha_k-1}(p_k - 1) = 14 .$$

Every number of the form $p_i - 1$ is either 1 or an even number, and therefore cannot be 7. In addition, $p_i - 1 \neq 14$, since 15 is not a prime. This means that one of the numbers $p_1^{\alpha-1}$ is 7. But then $p_i - 1 = 6$, and 14 is not divisible by 6, so $\varphi(m)$ is not 14 for any m.

Problem 4.6. The problem is easy to verify for $m = 1$. The next case to be considered is that in which m is some power of a prime; say, $m = p^\alpha$. For a number to be relatively prime to m, it is necessary and sufficient that it not be divisible by p. But among the numbers $0, 1, 2, \ldots, m - 1$, there are in all m/p numbers that are divisible by p. Consequently, there are in this list

$$m - \frac{m}{p} = m\left(1 - \frac{1}{p}\right) = p^\alpha\left(1 - \frac{1}{p}\right) = p^{\alpha-1}(p - 1) = \varphi(m)$$

numbers that are relatively prime to m.

We now note that for a to be relatively prime to m, it is necessary and sufficient that the remainder on division of a by m be relatively prime to m.

By what was established above, the number of possible remainders on division by $p_i^{\alpha_i}$ (that is, the number of numbers less than $p_i^{\alpha_i}$) that are relatively prime to $p_i^{\alpha_i}$ is equal to $\varphi(p_i^{\alpha_i})$. But, as was shown in problem 2.8, the congruence of two numbers modulo m follows from their congruence modulo $p_i^{\alpha_i}$ for all i, and conversely. In particular, for a number to be relatively prime to m, it is necessary and sufficient that it be relatively prime to each of the numbers $p_i^{\alpha_i}$.

Consequently, to each combination of remainders on division by $p_1^{\alpha_1}, p_2^{\alpha_2}, \ldots, p_k^{\alpha_k}$ that are relatively prime to the corresponding divisors, there corresponds exactly one remainder on division by m that is relatively prime to m. We now have only to note that the number of such combinations is equal to $\varphi(p_1^{\alpha_1})\varphi(p_2^{\alpha_2})\cdots\varphi(p_k^{\alpha_k}) = \varphi(m)$.

Problem 4.7. This is left to the reader.

Problem 4.8. $n^{13} - n = n(n^{12} - 1)$. But $n^{12} = n^{\varphi(13)} = n^{2\varphi(7)} = n^{3\varphi(5)} = n^{6\varphi(3)} = n^{12\varphi(2)}$. Therefore, either $p|n$ or $p|n^{12} - 1$ for $p = 2, 3, 5, 7,$ 13 (by Euler's theorem and the fact that in general, $x - 1$ divides $x^m - 1$ for any m); so, by theorem 1.6, $n^{13} - n$ is divisible by $2 \cdot 3 \cdot 5 \cdot 7 \cdot 13 = 2730$.

Problem 4.9. This is left for the reader to do independently.

Problem 4.10. Let d be the greatest common divisor of a and b. If c is not divisible by d, then it follows from theorem 1.6 that the equation $ax + by = c$ cannot be solved in integers. If c is divisible by d, however, then both sides of the equation may be divided by d, and we arrive at the case already considered; so the equation does have a solution in integers.

Problem 4.11. Let A and B be such that $aA + bB = 1$. Let us set

$$x_t = cA + bt, \qquad y_t = c\,\frac{1 - aA}{b} - at \,.$$

Then

$$ax_t + by_t = a(cA + bt) + b\left(c\,\frac{1 - aA}{b} - at\right)$$
$$= caA + abt + c(1 - aA) - abt = c \,,$$

and the pairs (x_t, y_t) are indeed solutions to our equation.

Problem 4.12. a. $x_t = 9 \cdot 5^5 + 7t = 28{,}125 + 7t,$

$$y_t = 9\,\frac{1 - 5^6}{7} - 5t = -20{,}088 - 5t \,.$$

Since the numbers 28,125 and 20,088 and the coefficients 7 and 5 are, so to speak, "approximately proportional," we may hope to obtain a representation of our solution using smaller numbers. Indeed, we may write

$$x_t = 6 + 7(t + 4017) \,, \qquad y_t = -3 - 5(t + 4017) \,,$$

or, setting $t + 4017 = t'$,

$$x_{t'} = 6 + 7t' \,, \qquad y_{t'} = -3 - 5t' \,.$$

b. Let us use the fact that the order of 25 modulo 13 is 2. We may write

$$x_t = 8 \cdot 25 + 13t = 200 + 13t,$$

$$y_t = 8\,\frac{1 - 25^2}{13} - 25t = -384 - 25t;$$

or, after simplifications,

$$x_{t'} = 5 + 13t', \qquad y_{t'} = -9 - 25t'.$$

Problem 4.13. Conditions (a), (b), and (c) are guaranteed automatically, and condition (d*) follows from theorem 4.6.

Problem 4.14. The criteria can be easily constructed using the general criterion and the following table:

m	17	19	27	29	31	49
k'	12 (or -5)	2	19 (or -8)	3	28 (or -3)	5

Problem 4.15. This is left for the reader.